The Cambridge Program for the Science Test

CAMBRIDGE Adult Education
Prentice Hall Regents, Englewood Cliffs, NJ 07632

The editor has made every effort to trace the ownership of all copyrighted material, and necessary permissions have been secured in most cases. Should there prove to be any question regarding the use of any material, regret is here expressed for such error. Upon notification of any such oversight, proper acknowledgment will be made in future editions.

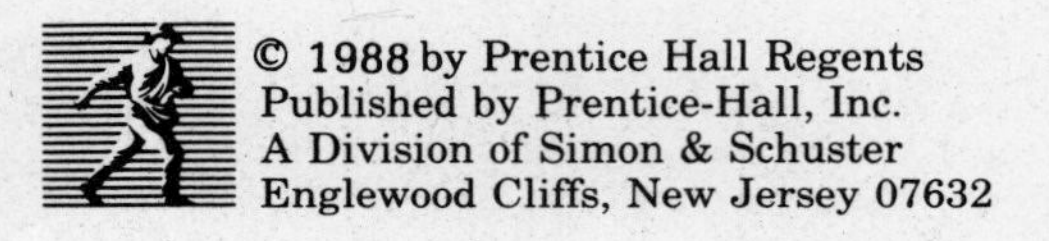

Published by Prentice-Hall, Inc.
A Division of Simon & Schuster
Englewood Cliffs, New Jersey 07632

Printed in the United States of America

10 9 8 7 6 5 4 3 2

ISBN 0-8428-8703-2

Prentice-Hall International (UK) Limited, *London*
Prentice-Hall of Australia Pty. Limited, *Sydney*
Prentice-Hall Canada Inc., *Toronto*
Prentice-Hall Hispanoamericana, S.A., *Mexico*
Prentice-Hall of India Private Limited, *New Delhi*
Prentice-Hall of Japan, Inc., *Tokyo*
Simon & Schuster Asia Pte. Ltd., *Singapore*
Editora Prentice-Hall do Brasil, Ltda., *Rio de Janeiro*

The Cambridge GED Program

The Cambridge GED Program

Writers

Gary Apple
Owen Boyle
Jesse Browner
Phyllis Cohen
Carole Deletiner
Randee Falk
Don Gerstein
Peter Guthrie
Alan Hines
Jeanne James
Lois Kasper
Rachel Kranz
Gloria Levine
Amy Litt
Dennis Mendyk
Rebecca Motil
Susan Muller
Marcia Mungenast
Thomas Repensek
Ada Rogers
Ann Rowe
Richard Rozakis
Elois Scott
Sally Stepanek
Steve Steurer
Carol Stone
Lynn Tiutczenko
Robin Usyak
Kenneth Uva
Shelley Uva
Tom Walz
Willa Wolcott
Patricia Wright-Stover
Karen Wunderman

Executive Editor

Jerry Long

Senior Editor

Timothy Foote

Project Editors

James Fina
Diane Maass

Subject Editors

Jim Bedell
Diane Engel
Randee Falk
Scott Gillam
Rebecca Motil
Thomas Repensek

Art and Design

Brian Crede Associates
Adele Scheff
Hal Keith

Contents

Practice

Simulation

Introduction

The following pages will introduce you to the Science Test of the GED Tests and to the organization of this book. You will read about ways you can use this book to your best advantage.

What Is the Science Test?

The Science Test of the GED Tests examines your ability to understand, use, analyze, and evaluate information from the life and physical sciences: biology, earth science, chemistry, and physics.

What Kind of Questions Are on the Test?

When you take the test, you will read passages and answer questions that test your understanding: You will be required to restate information, summarize ideas, or identify implications of the information given in the passage you read. You will also be tested on your ability to use and analyze what you understand. You may read a passage that describes a scientific phenomenon and then be asked to apply the information in a slightly different context. To do that you have to imagine how another situation might work, based on what you have been told about the first situation.

Some questions will test your ability to distinguish facts from opinions, to recognize assumptions, to tell the difference between a conclusion and its supporting statements, or to identify cause-and-effect relationships.

Finally, some passages may describe a scientific argument and then ask you to evaluate whether the data given seem to support the conclusion provided.

None of the questions will test only your prior knowledge of science. You will not be asked to give any formulas, to name any scientists, or to remember any dates. You will be asked only to demonstrate that you understand, can use, and can think critically about the major conceptual themes of science. All the questions on the Science Test are in a multiple-choice format.

What Are the Reading Passages Like?

The passages you will read are of varying lengths. For some short passages, you will have only one question to answer; for longer ones, you will usually have several. Sometimes you will be given graphic material instead of or along with written material. For example, you might be given a diagram of the world's time zones and then be asked to figure out what time it is in Tokyo when it is 6:00 P.M. in New York. Or you might be given a description of the behavior of gases in an enclosed space and then be asked to relate that behavior to the tendency of tires to expand on hot summer days.

Some of the passages on the Science Test will test your ability to relate scientific principles to life experience. For example, you might be given an advertisement and asked to distinguish claims about the product that are scientific facts from those that are opinions.

About half the passages will relate to biology; the other half will relate to the physical sciences: earth science, chemistry, and physics.

What You Will Find in This Book

This book gives you a four-step preparation for taking the Science Test. The four steps are as follows:

Step One: Prediction

In this first step, you will find the predictor test. This test is very much like the actual Science Test but is only half as long. Taking the predictor test will give you an idea of what the real GED will be like. By evaluating your performance on the predictor test, you will get a sense of your strengths and weaknesses. This information will help you to plan your studies accordingly.

Step Two: Instruction

The instruction section has two units. The first unit, Reading Strategies, can help you develop and sharpen your reading skills. You will learn useful strategies for approaching graphic materials such as maps, charts, graphs, tables, and diagrams. The first unit will also serve as a preparation for studying the second unit and taking the GED successfully.

Unit II, Foundations in Science, focuses on science material itself. In Unit II, you will read several lessons in each of the four content areas of science. These lessons develop the foundation of concepts and facts you should have when you take the GED. Even though *no* science question on the GED tests *only* your prior knowledge, *every* question requires you to draw on your general knowledge of science. The purpose of Unit II is to help you organize and add to the general knowledge about science that you already have.

Each lesson in Unit II ends with questions based on the material in the lesson. Even though the questions are usually not multiple-choice, they require you to think in the ways GED questions require you to think. The questions also give you the opportunity to practice writing. In completing this unit, you will be using the same kinds of thinking and reading skills that are required on the actual GED.

Both units of the instruction section are divided into chapters. Each chapter in Unit I covers a different aspect of reading skill. In Unit II there is a chapter for each type of science covered on the test: Biology, Earth Science, Chemistry, and Physics. Each chapter is divided into lessons.

Step Three: Practice

This section gives you valuable practice in answering the kinds of questions you will find on the actual Science Test. There are two separate types of practice activity in Step Three.

- **Practice Items** The Practice Items are GED-like questions grouped according to the content areas of science. For example, you will find items based on biology grouped together, items based on earth science grouped together, and so on. The practice items allow you to test your understanding of one field of science at a time.

- **Practice Test** The Practice Test is structured like the actual Science Test. In the Practice Test, the types of science vary from passage to passage, just as they do on the real test. This section gives you an opportunity to practice taking a test similar to the GED.

Each of the practice activities is made up of 66 items, the same number as are on the actual Science Test. You can use your results to track your progress and to give you an idea of how prepared you are to take the real test.

Step Four: Simulation

Finally, this book offers a simulated version of the Science Test. It is as similar to the real test as possible. The number of questions, their level of difficulty, and the way they are organized are the same. You will have the same amount of time to answer the questions as you will have on the actual test. Taking the Simulated Test will be useful preparation for taking the GED. It will help you find out how ready you are to take the real exam.

The Answer Key

At the back of this book, you will find a section called Answers and Explanations. The answer key contains the answers to all the questions in the Lesson Exercises, Chapter Quizzes, Unit Tests, Practice Items, the Practice Test, and the Simulated Test. The answer key is a valuable study tool: It not only tells you the right answer, but explains why each answer is right and points out the reading skill you need to answer each question successfully. You can benefit a great deal by consulting the answer key after completing an activity.

Using This Book

There are many ways to use this book. Whether you are working with an instructor or alone, you can use this book to prepare for the Science Test in the way that works best for you.

Take a Glance at the Table of Contents

Before doing anything else, look over the Table of Contents and get a feel for this book. You can compare the headings in the Table of Contents with the descriptions you have just read. You might also want to leaf through the book to see what each section looks like.

Take the Predictor Test

Next, you will probably want to take the Predictor Test. As the introduction to the test implies, there is more than one way to take this test. Decide which is best for you.

Your score on the Predictor Test will be very useful to you as you work with the rest of this book. It will point out your particular strengths and weaknesses, which can help you plan your course of study.

Beginning Your Instruction

After you have analyzed your strengths and weaknesses, you are ready to begin instruction. It would be best for you to work through Unit I before beginning Unit II, since the first unit focuses on the reading strategies you will use in the second unit.

At the beginning of Unit I you will find a Progress Chart. As you complete a lesson or chapter quiz, you can record your performance on the chart. The chart allows you to see your progress from lesson to lesson. If you feel you are not making enough progress, you can vary your method of studying or ask your teacher for help.

Unit II is organized according to the subject areas found on the test: biology, earth science, chemistry, and physics. You may want to work through the whole unit in order, completing the first chapter before going on to the second, or you may wish to work on two or more chapters at the same time. Again, there is a progress chart at the beginning of the unit to record your progress.

Using the Practice Section

When you complete a chapter in Unit II, you may proceed to the next chapter or you may wish to get some practice on the type of science you just studied. The Practice Items (pages 180–199) are grouped according to the types of science covered by the chapters of Unit II. You can test yourself immediately after each chapter, if you wish, or you can wait until you've finished all the chapters. You should take the Practice Test, however, only after you have finished both Units I and II.

Taking the Simulated Test

Finally, once you have completed the Instruction and Practice sections, you can take the Simulated Science Test. This will give you the most accurate assessment of how ready you are to take the actual test.

Try Your Best!

As you study the lessons and complete the activities and tests in this book, you should give it your best effort. To attain a passing score on the GED, you will probably need to get half or more of the items correct. To give yourself a margin for passing, try to maintain a score of at least 80 percent correct as you work through this book.

The Progress Charts will help you compare your work with this 80 percent figure. If you maintain 80 percent scores, you are probably working at a level that will allow you to do well on the GED.

Prediction

Introduction

Imagine that you are going to take the GED test today. How do you think you would do? In which areas would you perform best, and in which areas would you have the most trouble? The Predictor Test that follows can help you answer these questions. It is called a Predictor Test because your results on it can be used to show where your strengths and your weaknesses lie in relation to the actual Science Test of the GED.

The predictor test is like the actual GED Test in many ways. It will check your skills as you apply them to the kind of science passages you will find on the real test. The questions are like those on the actual test.

How to Take the Predictor Test

The Predictor Test will be most useful to you if you take it in a manner close to the way the actual test is given. If possible, you should complete it in one sitting with as little distraction as possible. So that you will have an accurate record of your performance, write your answers neatly on a sheet of paper or use an answer sheet provided by your teacher.

As you take the test, don't be discouraged if you find you are having difficulty with some (or even many) of the questions. The purpose of this test is to predict your overall performance on the GED and to locate your particular strengths and weaknesses. So relax—there will be plenty of opportunities to correct any weaknesses and retest yourself.

You may want to time yourself to see how long you take to complete the test. When you take the actual Science Test, you will be given 95 minutes. The Predictor Test is about half as long as the actual test, so if you finish within $47\frac{1}{2}$ minutes, you are right on target. At this stage, however, you shouldn't worry too much if it takes you longer.

When you are done, check your answer by using the answer key that begins on page 15. Put a check by each item you answered correctly.

How to Use Your Score

At the end of the test you will find a Performance Analysis Chart. Fill in the chart; it will help you find out which areas you are more comfortable with and which give you the most trouble.

As you begin each chapter in the book, you may want to refer back to the Performance Analysis Chart to see how well you did in that area of the Predictor Test.

PREDICTOR TEST

TIME: *$47\frac{1}{2}$ minutes*
Directions: *Choose the one best answer to each question.*

Item 1 is based on the following passage.

The mass of an object is constant, but the weight of an object is not. The weight of an object depends upon the force of gravity on the object. For example, a person who weighs 180 pounds on Earth would weigh only 30 pounds on the moon, because the force of gravity on the moon is about one-sixth as strong as the force of gravity on Earth.

1. An astronaut is orbiting Earth in a space vehicle. At this distance, the force of gravity is half as strong as it is on the surface of Earth. Which of the following statements is true?

(1) The astronaut's weight is half its value on the moon.
(2) The astronaut's weight is half its value on Earth.
(3) The astronaut's mass is half its value on Earth.
(4) The astronaut's weight is twice its value on Earth.
(5) The astronaut's mass and weight are half their value on Earth.

---

2. Tides are caused by the gravitational pull of the moon on Earth. At any given location on Earth, high tides and low tides will alternate about every six hours. However, because the moon rises about fifty minutes later each day, high tides and low tides begin about fifty minutes later each day.

If high tide at Sandy Hook Beach begins at 6 A.M. on June 23, at which of the following dates and times will high tide also begin?

(1) 12 noon on June 23
(2) 12:50 P.M. on June 23
(3) 6 A.M. on June 24
(4) 6:50 A.M. on June 24
(5) 6:50 A.M. on June 25

Items 3 to 7 are based on the following information.

Living things are classified according to certain characteristics. The largest classification group is called a kingdom. Listed below are the five kingdoms.

(1) plants: usually multicellular organisms incapable of moving from one place to another; able to make own food by photosynthesis, a process that converts carbon dioxide and water into carbohydrates while releasing oxygen into the atmosphere

(2) animals: multicellular organisms capable of moving from one place to another; have specialized tissues, organs, and organ systems; unable to make own food

(3) protists: unicellular organisms organized around a well-defined nucleus that directs the activities of the cell

(4) monerans: unicellular organisms lacking a well-defined nucleus

(5) fungi: unicellular or multicellular organisms unable to make own food due to lack of the green pigment chlorophyll; cannot move from place to place; have a distinct nucleus

Each of the following items describes an organism that can be classified in one of the five kingdoms. For each item, choose the one kingdom that best fits the characteristics of that organism. A kingdom *may* be used more than once in the following set of items.

3. Jellyfish move about in the ocean by using muscle tissue. Jellyfish have an umbrella-shaped, jellylike body and long tentacles. These tentacles protect the jellyfish and enable it to capture other organisms for food.

The jellyfish is a member of which kingdom?

(1) plants
(2) animals
(3) protists
(4) monerans
(5) fungi

4. In a closed environment, green algae removes carbon dioxide wastes and increases the supply of oxygen.

Green algae is a member of which kingdom?

(1) plants
(2) animals
(3) protists
(4) monerans
(5) fungi

5. Mold grows on certain foods such as bread. Mold consists largely of threadlike strands called hyphae. Long hyphae cover the surface of the bread, while short, rootlike hyphae grow into the bread. The shorter hyphae release enzymes that break down the bread so that it can be absorbed as food for the mold.

Mold is a member of which kingdom?

(1) plants
(2) animals
(3) protists
(4) monerans
(5) fungi

6. A bacterium is a single cell surrounded by a cell membrane, which controls the substances that enter and leave the cell. Unlike more advanced cells, bacteria do not have their hereditary material confined to a nucleus. Instead, the hereditary material is spread throughout the cytoplasm of the cell.

Bacteria are members of which kingdom?

(1) plants
(2) animals
(3) protists
(4) monerans
(5) fungi

7. The sea anemone looks like a brilliantly-colored underwater flower. But the "petals" are really poisonous tentacles designed to capture passing fish. Once the fish is stunned or killed, it is pulled into the sea anemone's mouth and quickly digested as food.

The sea anemone is a member of which kingdom?

(1) plants
(2) animals
(3) protists
(4) monerans
(5) fungi

Item 8 is based on the following figure.

Shown below is a diagram of water molecules (H_2O) as a solid (ice), liquid, and gas (water vapor).

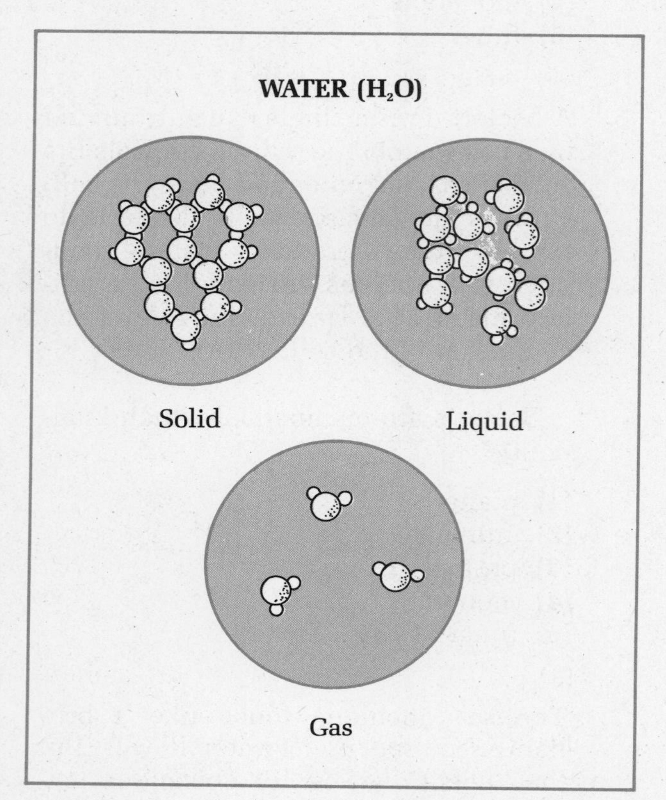

8. Which of the following statements best explains why ice floats in water?

(1) There are fewer atoms in a molecule of ice than there are in a molecule of water, so the ice weighs less.
(2) The molecules in ice are of a different, lighter type than those found in water.
(3) The rigid arrangement of ice allows it to float in water.
(4) The molecules in ice are buoyed up by water vapor rising from the surface of the water.
(5) The molecules in ice are farther apart than those in water, so the ice weighs less.

9. The movement of molecules of a substance into or out of a cell is called diffusion. Substances move from places where they are more concentrated to places where they are less concentrated.

If molecules of salt are moving from the inside to the outside of the cell, which of the following situations must be true?

(1) The concentration of salt is the same inside and outside the cell.
(2) The concentration of salt is greater outside the cell than inside the cell.
(3) The concentration of salt is greater inside the cell than outside the cell.
(4) The molecules of salt outside the cell are moving faster than the salt molecules inside the cell.
(5) The molecules of salt inside the cell are heavier than the salt molecules outside the cell.

Items 10 to 12 are based on the following diagram and passage.

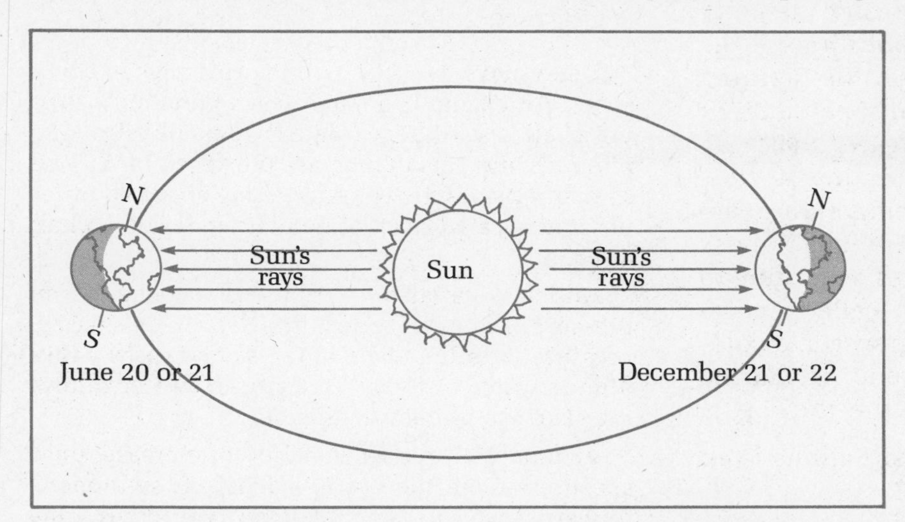

Due to the tilt of Earth's axis, the number of daylight hours in the northern and southern hemispheres is not constant. The hemisphere that leans toward the sun has longer days and shorter nights; the hemisphere that leans away from the sun has shorter days and longer nights. When summer begins in the northern hemisphere, the North Pole experiences 24 hours of daylight while the South Pole experiences 24 hours of darkness. The situation is reversed when winter begins in the northern hemisphere.

10. At which of the following locations will the number of daylight hours be longest on December 22?

(1) Arctic Circle, Greenland
(2) Denver, Colorado
(3) Mexico City, Mexico
(4) Sydney, Australia
(5) Roosevelt Island, Antarctica

11. A person wishes to live in an area where daylight hours are nearly constant all year. Where should this person live?

(1) near the North Pole
(2) near the South Pole
(3) near the equator
(4) halfway between the North Pole and the equator
(5) halfway between the South Pole and the equator

12. Which of the following situations would be true if Earth's axis were not tilted?

(1) The North Pole would always be in darkness.
(2) The South Pole would always be in darkness.
(3) Days would always be longer than nights in both hemispheres.
(4) Days would always be shorter than nights in both hemispheres.
(5) Days and nights would always be of equal length in both hemispheres.

13. Homeostasis is the ability of an organism to keep conditions inside its body constant even though conditions in its environment may change. Maintaining a relatively constant body temperature is an important part of homeostasis. Animals capable of maintaining a relatively constant body temperature are called warmblooded. Animals whose body temperature changes according to the environment are called coldblooded. In order to keep warm, a coldblooded animal must spend part of the day absorbing heat from the sun.

Which of the following situations is an example of homeostasis?

(1) An organism dies when the temperature reaches −80°F.
(2) A person sweats on a hot summer day to avoid overheating.
(3) A turtle lies in the sun in order to absorb heat.
(4) A crocodile becomes inactive at night when the temperature drops.
(5) A fish's body temperature changes when the surrounding water becomes warm or cold.

14. The normal gestation period, or length of pregnancy, for dogs is 61 days. A dog breeder wishes to have six-month-old puppies ready for sale by November 25. Puppies conceived on which of the following dates would be the right age for the sale?

(1) March 24
(2) April 25
(3) May 25
(4) August 26
(5) September 24

Items 15 to 18 refer to the following passage.

Athletes make use of scientific principles in every sport. For example, a major league pitcher throwing a ball toward home plate is using physics. The pitcher's arm acts to increase the speed of the ball so that it leaves his hand traveling about 90 miles per hour. It arrives at home plate in less than half a second.

An ice skater also makes use of science. When pressure is applied to ice, it melts. In ice skating, pressure from the blade of the skate causes a thin film of water to form. This film of water allows the skater to glide over the ice.

Swimmers have been able to increase their speeds through the use of science. A swimmer's forward speed is dependent on two forces. One force is the resistance of the water, which slows the swimmer down. The other force is that of the swimmer's arms and legs, which propel the swimmer forward. Swim coaches have applied this knowledge to devise strokes that increase the arm's push against the water. They also advise their teams to streamline the body's position in the water so as to decrease resistance.

The switch from bamboo poles to fiberglass poles in polevaulting is still another example of the use of science in sports. Bamboo poles break easily when bent, so polevaulters realized that they needed a more flexible and elastic pole. Fiberglass was used because it quickly returns to its original shape after bending. The result has been an increase in the world record for polevaulting by more than four feet.

15. In physics, a lever is a simple machine that consists of a bar that turns around a fixed point known as a fulcrum. When you push or pull one part of a lever, another part of it moves something. A lever is often used to increase the force used and thus to increase the speed of an object. The figure below depicts one type of lever.

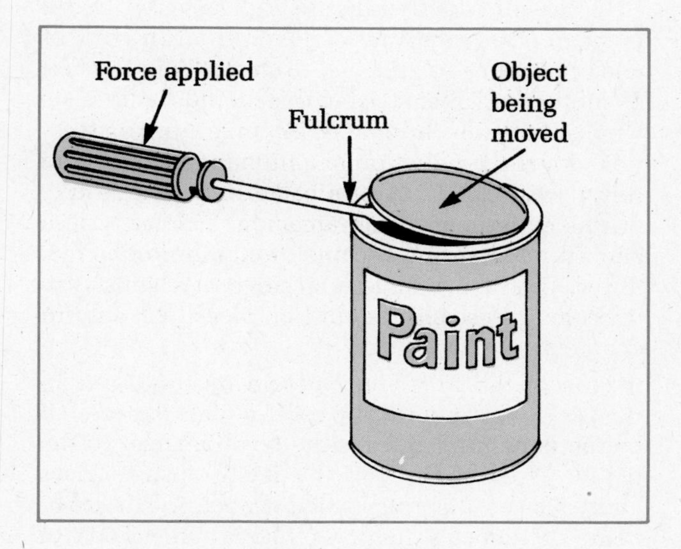

Which of the following devices described in the passage is NOT a lever?

(1) pitcher's arm
(2) ice skate's blade
(3) swimmer's arm
(4) swimmer's leg
(5) fiberglass pole

16. At very cold temperatures (below −8°F), it is not possible to ice skate. This is because

(1) ice changes to a solid at these temperatures
(2) ice changes to a liquid at these temperatures
(3) ice changes to a gas at these temperatures
(4) it becomes too warm for pressure to melt the ice
(5) it becomes too cold for pressure to melt the ice

17. Which of the following statements is NOT evidence that fiberglass is more flexible and elastic than bamboo?

(1) Bamboo poles break more often than fiberglass poles do.
(2) Fiberglass returns to its original shape after bending more quickly than does bamboo.
(3) Since fiberglass poles were introduced, the world record for pole-vaulting has increased by more than four feet.
(4) The current world champion pole-vaulter uses a fiberglass pole, not a bamboo pole.
(5) After repeated use, fiberglass poles retain their original shape better than do bamboo poles.

18. For which of the following sports is resistance of the medium (air or water) in which the sport is conducted the biggest problem?

(1) running
(2) walking
(3) bicycling
(4) football
(5) sailing

19. A fire extinguisher contains liquid carbon dioxide under pressure. When you use the extinguisher, the pressure drops and the carbon dioxide rushes out as a gas. The escaping gas becomes very cold, forming a solid known as dry ice. Which of the following statements best explains why the carbon dioxide gas becomes dry ice?

(1) Carbon dioxide is a solid at room temperature.
(2) The outside air reacts with the carbon dioxide gas to form dry ice.
(3) Moisture in the air reacts with the carbon dioxide gas to form dry ice.
(4) The liquid carbon dioxide expands to a gas so quickly that enough energy is used to cool it into a solid.
(5) The liquid carbon dioxide is really a solid compressed into the liquid state while in the extinguisher.

20. A stimulus is any action or agent that causes an activity in an organism. This activity is called a response. A response is an action or movement on the part of the organism. Blinking is a response to the stimulus of sudden movement in front of the eyes. Closing its petals is the lily's response to the absence of light.

A botanist wishes to make a lopsided plant grow straight. Which of the following would be an appropriate stimulus to obtain this response?

(1) placing a light directly above the plant
(2) propping the plant up with a stick
(3) using wire mesh to hold the plant in place
(4) altering the genetic material in the plant
(5) cutting off the crooked part of the plant

Items 21 to 25 are based on the following passage.

Imagine an organism so small that it can be seen only through a microscope. Then imagine that this same organism produces barrels full of useful human hormones that are extremely scarce. This is what is happening in laboratories today as the result of a revolutionary technique known as genetic engineering.

In the mid-1970s, scientists discovered that a piece of DNA could be removed from the cell of one organism and attached to the DNA in the cell of another organism. Using this method, scientists were able to take human DNA containing instructions for the production of a human hormone and link it with the DNA of a bacterium. The results of this experiment were amazing. The bacterium and all its offspring became human hormone factories, producing huge quantities of a substance once so scarce that it could be measured only in droplets.

One of the first hormones to be produced in this way was insulin. Insulin, which is secreted by the pancreas, controls the level of sugar in the blood. Without insulin, the level of sugar rises and causes a disorder called diabetes. The treatment of diabetes involves regular injections of insulin. Before the production of human insulin by bacteria, diabetics had to use insulin obtained from pigs or cows. This animal insulin sometimes produced allergic reactions. It was also expensive and hard to obtain. Scientists predict that the bacteria-produced insulin will be plentiful and inexpensive and will not cause allergies.

Another hormone produced by genetically engineered bacteria is interferon. Interferon is produced in human cells to help fight viruses. Scientists believed that interferon could be used to treat many viral infections, including cancer. Until the production of interferon by bacteria, however, there was never enough interferon available to test these ideas.

21. According to the passage, genetic engineering involves which of the following processes?

(1) exchanging the DNA of one organism with the DNA of another organism
(2) removing DNA from bacteria and placing it in humans
(3) injecting bacteria into human DNA
(4) combining the DNA of one organism with the DNA of another organism
(5) separating DNA into parts and placing each part in a different organism

22. It can be inferred from the passage that scientists were eager to create hormone-producing bacteria because

(1) they wanted to learn more about the behavior of bacteria
(2) they wanted to compare the hormones produced by bacteria with the hormones obtained from pigs and cows
(3) they wanted to find a way to increase the available supply of human hormones
(4) they wanted to alter the structure of human DNA
(5) they wanted to find a cure for cancer

23. Based on the information provided, which of the following developments is likely to occur as a result of genetic engineering?

(1) Treatment of disorders such as diabetes will become less expensive.
(2) Fewer people will develop disorders caused by hormone deficiencies.
(3) Fewer people will be allergic to insulin obtained from cows or pigs.
(4) Less research involving human hormones will be needed.
(5) More disease-producing bacteria will be present in the environment.

24. A hypothesis is an "educated guess" that has yet to be tested by experiment. Based on the information provided, which of the following statements represents a hypothesis, rather than a statement of fact?

(1) Bacteria can produce large quantities of human hormones.
(2) The offspring of a genetically engineered bacterium will produce the same hormone as the parent bacterium.
(3) Interferon can be used to treat viral infections.
(4) Human hormones have often been difficult to obtain.
(5) Insulin is necessary to control the level of sugar in human blood.

25. Experiments are now being carried out to determine whether interferon can be used to cure cancer. Based on the information provided, which of the following statements best explains why these experiments are possible?

(1) More cancer patients are willing to try interferon treatments.
(2) The government is now providing funds for interferon research.
(3) More physicians are recommending treatment with interferon.
(4) Interferon is now available in larger quantities.
(5) Interferon is now less expensive than other available cancer treatments.

Item 26 refers to the following figure and text.

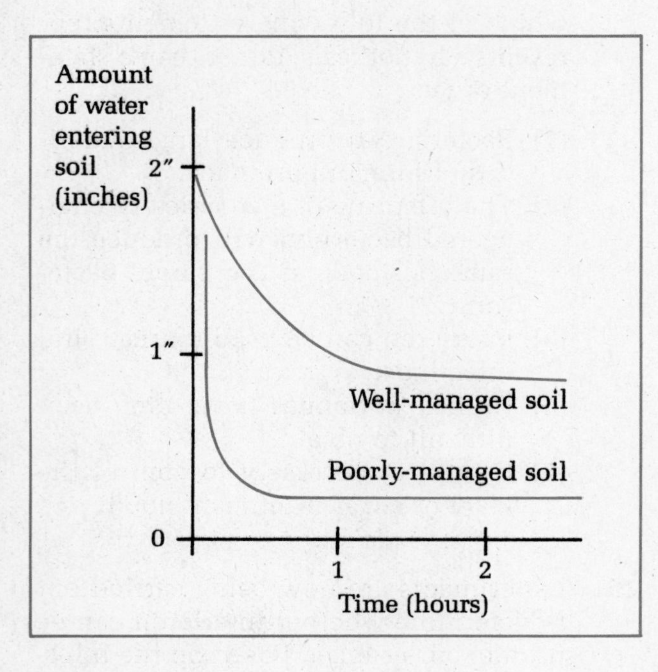

The graph shows the different rates at which soil absorbs water. Well-managed soil has a covering of vegetation or grasses. Poorly-managed soil tends to be bare and compacted by animals or vehicles.

26. Heavy grazing by cattle reduces the amount of rainwater soil can absorb. Based on the information provided, which of the following statements would justify this conclusion?

(1) Poorly managed soil cannot absorb as much water as well-managed soil.
(2) Well-managed soil cannot absorb as much water as poorly managed soil.
(3) Cattle-grazing land receives less rain than land used for other purposes.
(4) Allowing cattle to overgraze is an example of poor soil management.
(5) Raising cattle involves a great deal of human, animal, and vehicle traffic, which tends to compact the soil.

27. The relationship of power, voltage, and current is expressed by the formula, Power = Voltage × Current. A certain household lamp operates at 125 volts. If the maximum current the lamp can carry safely is 0.8 amps, which of the following light bulbs would prove hazardous if used in the lamp?

(1) 30 watts
(2) 60 watts
(3) 75 watts
(4) 100 watts
(5) 150 watts

Items 28 and 29 are based on the following graph.

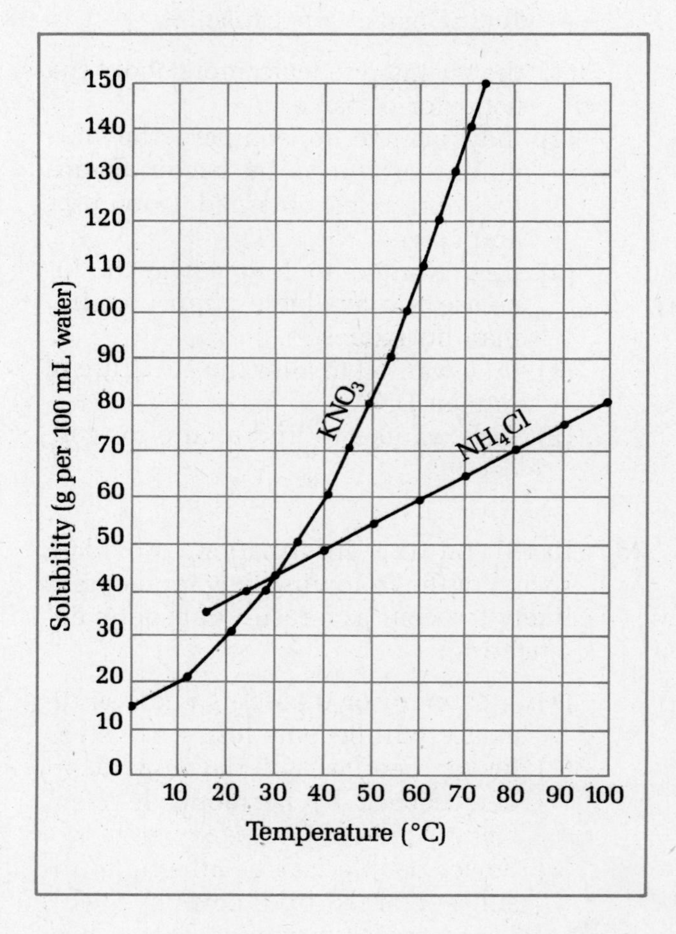

In the above graph, every point on each curve represents the number of grams of solute that will dissolve in 100 milliliters of water at a given temperature.

28. How many grams of potassium nitrate (KNO_3) will dissolve in one liter of water at 60°C (one liter = 1000 milliliters)?

(1) 40
(2) 110
(3) 125
(4) 600
(5) 1100

29. Based on the graph, at what temperature would you predict that 90 grams of NH_4Cl would dissolve in 100 mL of water?

(1) 50°C
(2) 100°C
(3) 120°C
(4) 150°C
(5) 200°C

---

30. A habit is an activity that becomes automatic and involuntary through constant repetition. At first, habitual behavior may require thought and effort, but eventually it can be done "without thinking." For example, the mechanics of handwriting eventually become habit. Riding a bicycle is another example of a habit.

According to the information provided, which of the following actions would be considered a habit?

(1) tying shoelaces
(2) jumping when startled by a loud noise
(3) writing a computer program
(4) going to the movies every Saturday night
(5) drinking when thirsty

Item 31 refers to the following figure.

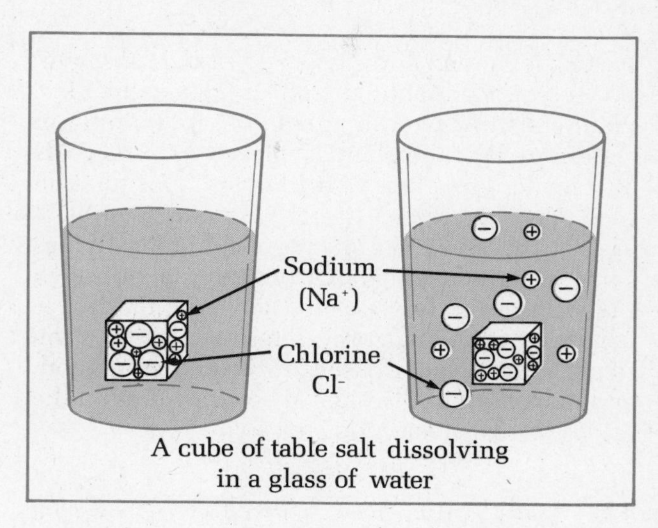

A cube of table salt dissolving in a glass of water

31. An ion is an atom having an electrical charge. The electrical charge may be either positive (+) or negative (−). Table salt, or sodium chloride, is considered an ionic compound. If table salt is a typical ionic compound, which of the following statements is true, based on information in the figure?

(1) Ionic compounds must reach a certain temperature before they will dissolve in water.
(2) When ionic compounds dissolve in water, they come apart as ions rather than neutral atoms (those with neither a positive nor a negative charge).
(3) Ionic compounds produce an electric current as they dissolve in water.
(4) Ionic compounds in water dissolve into positively charged ions only.
(5) Ionic compounds in water dissolve into negatively charged ions only.

Items 32 and 33 refer to the following passage.

Plants have four basic types of tissues—growth, protective, conducting, and fundamental. The plant cells involved in growth are found only in certain parts of the plant, such as the tips of roots and stems. The plant's protective tissues form an outer layer to reduce water loss and to protect against diseases. The plant's conducting tissues carry water and minerals from the roots through the stem to the leaves. Food made by the leaves is carried by similar tissues to other parts of the plant. Fundamental tissues make and store food, and help support the plant. They are found in the leaves, fruits, stems, and roots.

32. Which of the following processes would peeling the bark off a tree interfere with most directly?

(1) transportation of water to the leaves
(2) transportation of food to the roots
(3) protection against water loss
(4) stem growth
(5) food production

33. Pinching back a plant to slow growth without harming the plant would be most effective at which point?

(1) tip of the roots
(2) tip of the leaves
(3) middle of the leaves
(4) tip of the stem
(5) middle of the stem

Answers are on page 15.

Answers and Explanations for the Predictor Test

1. (2) *Application/Physics.* Since the weight of an object is directly proportional to the force of gravity on that object, the astronaut would weigh half of what he would weigh on Earth. The mass would not be affected.

2. (4) *Comprehension/Earth Science.* Since high tide begins about 50 minutes later each day, high tide would begin at 6:50 A.M. on June 24. Choices (1) and (2) are close to the time that low tide would begin. Choice (5) is wrong because by June 25, high tide would begin at 7:40 A.M.

3. (2) *Application/Biology.* Since the jellyfish can move from place to place, has specialized muscle tissue, and is dependent upon other organisms for food, it is a member of the animal kingdom.

4. (1) *Application/Biology.* The use of carbon dioxide and the release of oxygen indicate the process of photosynthesis. Therefore green algae is a plant.

5. (5) *Application/Biology.* Since mold is a multicellular organism that cannot make its own food or move from place to place, it belongs to the fungi kingdom.

6. (4) *Application/Biology.* Since the bacteria cell lacks a well-developed nucleus, it is classified as a moneran.

7. (2) *Application/Biology.* Because the sea anemone cannot produce its own food (even though it looks like a plant) and has specialized organs for digestion and poisoning prey, it is a member of the animal kingdom.

8. (5) *Analysis/Chemistry.* The figure shows the same molecule, H_2O, in three different phases. The solid phase shows the molecules are farther apart than in the liquid phase, so ice weighs less than the same amount of water. There is no evidence in the figure for choices (2), (3), and (4). Choice (1) is false according to the figure.

9. (3) *Comprehension/Chemistry.* Since substances move from areas of greater concentration to areas of lesser concentration, salt must be more highly concentrated inside the cell. Choices (4) and (5) are irrelevant to the information provided.

10. (5) *Comprehension/Earth Science.* On December 22 the southern hemisphere is tilted toward the sun. Although Sydney, Australia, is also in the southern hemisphere, Roosevelt Island, Antarctica, is closer to the South Pole.

11. (3) *Application/Earth Science.* Since the equator is halfway between the two poles, its orientation toward the sun does not change no matter which way Earth is tilted. The hours of daylight and darkness are equal at the equator.

12. (5) *Evaluation/Earth Science.* If Earth's axis were not tilted, the length of day and night would depend only on Earth's rotation about its axis. Locations in both hemispheres would experience 12 hours of daylight when facing the sun and 12 hours of darkness when facing away from the sun.

13. (2) *Application/Biology.* Sweat removes heat from the body, enabling the person to maintain a constant body temperature. Choice (3) is incorrect because the turtle is absorbing heat from the sun, which is an external source; the turtle will get cold again once the sun goes down. Choices (1), (4), and (5) are examples of an organism not being able to keep its body temperature constant in a changing environment.

14. (1) *Application/Biology.* Dogs conceived on March 24 would be born on May 23 and would be six months old by November 23. If you answered choice (3), you probably forgot to consider the 61-day gestation period. If you answered choice (5), you probably forgot to consider the six-month growth period.

15. (2) *Application/Physics.* A pitcher's arm has levers that pivot around the shoulder and elbow. A swimmer's leg pivots around the hip to deliver a push forward in the water. A fiberglass pole pivots at the point where it touches the ground, delivering a push into the air. An ice skate blade is not a lever because you don't push or pull at one part of it to move another part of it. However, the ice skater's leg is a lever.

16. (5) *Analysis/Physics.* Ice skates work because pressure from the blade of the skate causes a thin film of water to form. If ice skating becomes impossible below a certain temperature, it must be too cold for the film of water to form no matter what the pressure.

17. (4) *Evaluation/Physics.* The world champion's choice of pole may be a matter of preference based on something other than fiberglass's elastic properties, so this is not in itself evidence that fiberglass is more flexible and elastic than bamboo.

18. (5) *Comprehension/Physics.* Water offers greater resistance to forward movement than does air, so any sport conducted in water would find resistance a problem.

19. (4) *Analysis/Physics.* When a material changes phase from a solid to a liquid, or liquid to gas, energy is required. However, when a liquid under pressure is suddenly released from pressure, energy is not added, so the phase change requires a use of heat as energy to take place. The gas molecules lose the energy necessary to remain a gas, becoming a solid.

20. (1) *Application/Biology.* Placing a light directly above the plant would cause the plant to react by growing up toward the light. Choices (2), (3), and (5) represent things done externally *to* the plant, not a reaction on the part of the plant. Choice (4) is wrong because altering genetic material would change future plants but not this plant.

21. (4) *Comprehension/Biology.* A piece of human DNA is attached to the DNA of a bacterium, resulting in a piece of combined DNA.

22. (3) *Analysis/Biology.* Both medical treatment and research have been hampered by insufficient supplies of human hormones. No other motivation for creating genetically engineered bacteria is discussed.

23. (1) *Analysis/Biology.* The passage implies that in the past, a hormone such as insulin was expensive because it was difficult to obtain and in short supply. It is logical to assume that the huge, inexpensive supply of hormone obtainable from bacteria will significantly reduce the cost of treatment for disorders such as diabetes.

24. (3) *Analysis/Biology.* According to the passage, scientists believe that interferon can be used to treat viral infections, but no experiments are cited to prove or disprove this hypothesis.

25. (4) *Evaluation/Biology.* Scientists have been unable to carry out experiments involving interferon because there has not been enough interferon available. It can be assumed that if such experiments are now being carried out, the increased supply of interferon brought about by genetic engineering has made the experiments possible.

26. (1) *Evaluation/Earth Science.* Poorly managed soil lacks a covering of vegetation and is compacted by animals or vehicles. Excess grazing is thus an example of poorly managed soil. The graph shows that poorly managed soil absorbs less water per hour than does well-managed soil, leading to the stated conclusion. Choice (2) is false. Choice (3) is improbable and lacks justification, according to the information provided. Choices (4) and (5) are true, but do not lead directly to the conclusion.

27. (5) *Analysis/Physics.* 150 watts = 125 volts × 1.2 amps. A 150-watt bulb would require a current of 1.2 amps—much more than 0.8 amps.

28. (5) *Application/Chemistry.* At a temperature of 60°C, 110 grams of KNO_3 will dissolve in 100 mL of water. Since one liter is equal to 1000 mL, the answer is 110 × 10 or 1100 grams. Choice (2) is incorrect because one liter of water is not the same as 100 mL of water.

29. (3) *Analysis/Chemistry.* Since the solubility curve for NH_4Cl is linear, you can extend the line to see where it would reach 90 grams. It crosses 90 grams when the temperature is 120°C.

30. (1) *Comprehension/Biology.* Tying shoelaces becomes habit in much the same way as the mechanics of handwriting. Jumping at the sound of a loud noise is an automatic response that is not learned. Drinking water when thirsty is an inborn or instinctive behavior. While going to the movies every week might be called a "habit" according to everyday usage of the word, it is not a habit according to the definition given.

31. (2) *Analysis/Chemistry.* The figure shows that the sodium and chloride ions still retain their charges after they are dissolved in water, so they are considered ions by the definition given. Choice (1) is incorrect because there is no indication of a temperature change. Choice (3) is incorrect because the ions haven't produced an electric current. Choices (4) and (5) are incorrect because both positively charged ions and negatively charged ions are in the water.

32. (3) *Analysis/Biology.* The bark is the outer layer of the tree. The outer layers of a plant serve to protect against water loss.

33. (4) *Analysis/Biology.* The tips of the roots and stems are where growth cells are found, eliminating choices (2), (3), and (5). Pinching back the roots would not be as effective as pinching back the stem tip, because the roots are also important in obtaining water and minerals for the plant. Pinching back the roots might harm the plant, as well as slowing growth.

PREDICTOR TEST
Performance Analysis Chart

Directions: Circle the number of each item that you got correct on the Predictor Test. Count how many items you got correct in each row; count how many items you got correct in each column. Write the amount correct per row and column as the numerator in the fraction in the appropriate "Total Correct" box. (The denominators represent the total number of items in the row or column.) Write the grand total correct over the denominator, **33,** at the lower right corner of the chart. (For example, if you got 28 items correct, write *28* so that the fraction reads *28*/**33.**) Item numbers in color represent items based on graphic material.

Item Type	*Biology (page 83)*	*Earth Science (page 106)*	*Chemistry (page 123)*	*Physics (page 142)*	TOTAL CORRECT
Comprehension (page 31)	21, 30	2, 10	9	18	**/6**
Application (page 39)	3, 4, 5, 6, 7, 13, 14, 20	11	28	1, 15	**/12**
Analysis (page 46)	22, 23, 24, 32, 33		8, 29, 31	16, 19, 27	**/11**
Evaluation (page 57)	25	12, 26		17	**/4**
TOTAL CORRECT	**/16**	**/5**	**/5**	**/7**	**/33**

The page numbers in parentheses indicate where in this book you can find the beginning of specific instruction about the various fields of science and about the types of questions you encountered in the Predictor Test.

Instruction

Introduction

This section of the book contains lessons and exercises that can help you learn the things you need to know to pass the GED Science Test.

The instruction section is divided into two units. Unit I, *Reading Strategies*, will help you to improve your skills at reading effectively. Unit II, *Foundations in Science*, can help you organize and increase your general knowledge of science and show you how to apply your reading skills to information about science presented in either narrative or graphic format.

The units are divided into chapters, which are in turn further divided. There are many exercises and quizzes, so you will have several opportunities to apply and test your understanding of the material you study. A progress chart at the beginning of each unit will make it easy for you to keep track of your work and record your performance on each lesson.

UNIT

I

Reading Strategies

In this unit, you will learn important strategies for reading. When you take the Science Test, you will be asked to read passages and then answer questions based on those passages. The better you understand the passages you read, the better you will perform on the test. The strategies you study in this unit will help you to better understand the passages you read.

Questions on the Science Test ask you to demonstrate that you understand what you read and that you can read critically. This unit will help you develop strategies for (1) determining how a question based on a science passage requires you to think and (2) answering questions effectively.

Unit I ends with a Unit Test that will check your understanding of and ability to use the reading strategies. You should complete Unit I before going on to Unit II.

UNIT I PROGRESS CHART
Reading Strategies

Directions: Use the following chart to keep track of your work. When you complete a lesson and have checked your answers to the items in the exercise, circle the number of questions you answered correctly. When you complete a Chapter Quiz, record your score on the appropriate line. The numbers in color represent a score of 80% or better.

Lesson	Page	
		CHAPTER 1: **A Strategy for Reading**
1	23	Previewing Test Items **1 2 3**
2	25	Questioning As You Read Test Items **1 2 3**
3	27	The Four Levels of Questions **1 2 3**
	29	Chapter 1 Quiz **1 2 3 4 5 6 7 8 9 10**
		CHAPTER 2: **Comprehension**
1	31	Restatement **1 2 3**
2	34	Summarizing Ideas **1 2 3**
3	36	Identifying Implications **1 2 3**
	37	Chapter 2 Quiz **1 2 3 4 5 6 7 8 9 10**
		CHAPTER 3: **Application**
1	39	Application from Passages and Graphics **1 2 3**
2	41	Application from a List of Categories **1 2 3**
	44	Chapter 3 Quiz **1 2 3 4 5 6 7 8 9 10**

Lesson	Page	
		CHAPTER 4: **Analysis**
1	46	Facts versus Opinions and Hypotheses **1 2 3**
2	48	Identifying Unstated Assumptions **1 2 3**
3	51	Conclusions versus Supporting Statements **1 2 3**
4	53	Identifying Cause-and-Effect Relationships **1 2 3**
	54	Chapter 4 Quiz **1 2 3 4 5 6 7 8 9 10**
		CHAPTER 5: **Evaluation**
1	57	Adequacy and Appropriateness of Information **1 2 3**
2	59	Effect of Values and Beliefs on Decisionmaking **1 2 3**
3	60	Accurary of Facts Based on Documentation **1 2 3**
4	62	Indicate Logical Fallacies in Arguments **1 2 3**
	64	Chapter 5 Quiz **1 2 3 4 5 6 7 8 9 10**

Chapter 1

A Strategy for Reading

Objective

In this chapter, you will

- Preview material you are about to read
- Question material as you read
- Do previewing and questioning while reading items based on life science and physical science material
- See how previewing and questioning will help you answer items on the Science section of the GED

Lesson 1 Previewing Test Items

Why should you preview written material before you read it?

A TV preview is a brief look at some of the high points of a show. It is designed to give you an idea of what the show will be about so that you will want to watch the entire program.

You can also preview written material by quickly looking it over. Previewing helps motivate you to read further. It also helps you make a mental outline of how information is organized. Both the motivation and the mental outline will make it easier for you to do the reading.

You probably already preview many things you read. For example, when you open a magazine, you probably flip through it. From the table of contents, the titles of articles, and the illustrations, you get an idea of what the magazine contains.

Steps in Previewing

1. Note where directions begin.
2. Note where the last item based on the passage or graphic ends.
3. Look at the form of the information presented. Is there a graph? a table? Is there a brief passage? a longer passage? one item? several items? Is there a set of items, all containing the same five choices?
4. Notice any headings, captions, titles, or labels.
5. Observe any words that "jump off the page" because they are set off by italics, dark print, underlining, extra spacing, capitalization, or repetition.

Remember, your preview should take only a few seconds. You are just trying to get an idea of the overall form of the information and an idea of the general topic.

Use what you have just learned to preview the following item.

Item 1 is based on the following table.

Energy Use by Home Appliances (in kilowatt-hours used annually)

Appliance	kWh	Appliance	kWh
Food Preparation		**Laundry**	
Blender	15	Clothes dryer	993
Dishwasher	363	Iron	144
Microwave oven	190	Washing machine	103
Range with oven	1,175	Water heater	4,219
Toaster	39	**Home Entertainment**	
Food Preservation		Radio	86
Frostless freezer	1,761	Stereo	109
Frostless refrigerator	1,217	Color television	440

From John Fowler, *The Energy-Environment Source Book.* National Science Teachers Association, 1979.

1. Which two appliances use the most energy?

(1) Range with oven and frostless freezer
(2) Frostless freezer and frostless refrigerator
(3) Frostless refrigerator and water heater
(4) Water heater and frostless freezer
(5) Range with oven and frostless refrigerator

In your preview, note that:

- You begin with the words in italics, "*Item 1* is . . ."
- Choice **(5)** of Item 1 marks the end of the material you preview when approaching this table. Can you see at a glance how much material you will need to look at in order to answer this item? Since there is only one item, it wouldn't make sense to spend too much time on the table.
- Words and numbers in the table are arranged into columns and rows. This arrangement makes it easier for you to find, understand, and compare the figures.
- The title at the top of the table tells you that the table is concerned with how much energy is needed to run the appliances in a home.
- The appliances are organized into four capitalized headings.
- The energy used is expressed as kilowatt-hours used yearly.
- The table is cut into two halves to save space.

By previewing the table for just a few moments, you have picked up a lot of information about its form and contents.

Previewing is the first step in an effective reading strategy. By taking a quick look at how material is arranged on the page, you prepare yourself for the active reading that comes next.

Lesson 1 Exercise

Items 1 to 3, below, are based on the following passage and its item.

An advertisement for popular fast-food French fries states:

A. Fried in 100% vegetable oil
B. No hamburger's complete without them
C. No added salt
D. Tastier than fries from Quick Burger Restaurant

Which of the above statements is most likely based on fact rather than on opinion?

(1) A only
(2) B only
(3) A and B only
(4) A and C only
(5) B and D only

1. A preview of the written material indicates that the information is presented in a paragraph/graph/list. (Choose one.)
2. TRUE or FALSE: A preview of the item shows that you will need to examine each choice carefully before arriving at your answer. Explain your answer.
3. Explain why it is better to preview the material first rather than simply reading it from start to finish.

Answers are on page 253.

Lesson 2 Questioning as You Read Test Items

Why should you use questioning as you read?

As you watch TV previews, you probably ask yourself, What is the story behind the key scenes shown? Then, when the title of the episode appears on the screen, you may ask, How does the title fit together with the pieces of the show I have just seen? You may even make some predictions about the story. Then you watch to see what actually happens.

In a similar way, you can use the questioning process when reading material you have previewed. As you ask each question, you create a purpose for reading further.

After previewing material on the Science section of the GED, apply questioning as you read.

1. Ask a question that you think might be answered in the passage.
2. Predict an answer.
3. Read to see whether your prediction is right.

What steps would you take to question the passage and item, below, after previewing them?

Many common household materials can produce toxic (poisonous) fumes. Radon gas and formaldehyde fumes are being found in an increasing number of homes. Radon gas can come from certain concrete foundations. Formaldehyde fumes are sometimes given off by permanent-press fabrics, plywood, drapes, and foam insulation.

According to the passage, which of the following statements is true about houses with concrete foundations?

(1) They all contain radon gas.
(2) They all contain formaldehyde fumes.
(3) Some contain radon gas.
(4) They contain more formaldehyde fumes than radon gas.
(5) They contain more radon gas than formaldehyde fumes.

After previewing the material, apply the questioning process:

1. Read and reword the question that the item asks. What can you assume about houses with concrete foundations?
2. Before you read the passage, make your own prediction. You may have noticed the key words *toxic fumes* in your preview. You might predict that concrete foundations make houses safer by keeping toxic fumes from coming into the house.
3. Read the passage. You find that your prediction is wrong. Concrete foundations sometimes actually *produce* toxic fumes.

Whether your predictions turn out to be right or wrong is not really important. By asking questions, you get involved in your reading. Ask yourself as many or as few questions as you need to keep yourself interested in the reading.

Lesson 2 Exercise

1. The questions you ask as you read do not all have to come from the test questions. For example, after reading the first sentence of the passage about toxic gases in the home, what question might you ask yourself?
2. Suppose you are trying to convince a friend of the value of questioning as you read. Your friend says, "When you stop to make up questions, doesn't that slow down your reading?" You answer, "____________."
3. Suppose you come to a diagram on the GED with the title "Some Occupations Related to Interest and Ability in Biology." What question might you ask yourself as you look more closely at the diagram?

Answers are on page 253.

Lesson 3 The Four Levels of Questions

What are the four levels of questions on the Science test, and why should you learn to identify them?

Just as questions you face in real life vary in difficulty, so do the items on the GED. There are four levels of questions on the GED: comprehension, application, analysis, and evaluation.

Comprehension questions, those at the lowest and often easiest level, require you to show that you understand the passage or graphic. For example, suppose you are presented with the following statements:

> Fatigued, but determined to stay awake, Gwen settled into the comfortable seat with her popcorn. The next thing she knew, the lights were on and the audience was leaving.

A comprehension question might be: What does the word *fatigued* mean? Your answer will show whether you understand the definition of *fatigued*. Even if you had never heard the word before, there are enough clues in the passage to help you guess that it means "tired."

Application questions require you to understand the information you are given and to apply that understanding to a new situation. An application question based on the statements about Gwen might be: What do you think would have happened if Gwen had decided to watch TV instead of going out? Even though the passage does not say anything about TV, you can use what you are told about Gwen's fatigue to figure out that she probably would have fallen asleep in front of the TV.

Analysis questions require that you not only understand the ideas, but also think about relationships between them, such as cause and effect. An analysis question might be: What caused the audience to leave? Using what you are told about Gwen's situation plus your own experience with theaters, you can speculate that the movie was probably over.

Evaluation questions require that you judge the accuracy of information based on standards you are given. An evaluation question related to the passage might be:

> Which of the following statements is better supported by the evidence?
>
> (1) Gwen missed the movie.
>
> (2) Gwen enjoyed the movie.

The answer key in this book labels the level of every question. You can practice identifying levels by predicting the level of a given question, then checking the answer key.

Abilities Measured by Each of the Four Levels of Questions

Each of the four levels of questions places different demands on your thinking. Once you recognize these demands, you will know how to read the material so that you will be able to answer the item successfully.

Questions at the comprehension level measure the ability to

- restate information
- summarize ideas
- identify implications

Questions at the application level measure the ability to

- use given or remembered ideas in a new situation

Questions at the analysis level measure the ability to

- distinguish facts from hypotheses or opinions
- recognize unstated assumptions
- distinguish a conclusion from supporting statements
- identify cause-and-effect relationships

Questions at the evaluation level measure the ability to

- assess the adequacy of data to support conclusions
- recognize the role values play in beliefs
- assess the accuracy of facts as determined by proof
- indicate logical fallacies in arguments

Lesson 3 Exercise

1. Based on the information in this lesson, which of the following statements is NOT true?
   (1) Evaluation questions will be asked after graphics (tables, charts, etc.) as well as after passages.
   (2) Before you can answer an analysis question, you have to be able to comprehend the material.
   (3) You can often identify the level of a question by its wording.
   (4) Comprehension questions are generally easier than higher level questions.
   (5) Application items ask you to summarize the information presented.

2. Make up one question at each of the four levels based on the following information.

   If you are buying a tree or shrub, look for one that appears fresh. Wilting usually indicates that the plant is suffering from lack of water. It's very hard on a plant to allow it to dry out just when it is starting to grow.

3. The main advantage of being able to identify the level of an item that you meet on the Science section of the GED is that ______.

Answers are on page 253.

Chapter 1 Quiz

1. Change the underlined words to make the following a true statement: To preview a chart or diagram, <u>you should look carefully at every word from top to bottom.</u>

2. There are advantages to predicting an answer to a test item before you actually read the choices. Can you think of two advantages?

Items 3 and 4 are based on a preview of the following passage.

Since energy is lost when there is friction (rubbing),

(1) energy is gained when there is no friction.
(2) energy can be saved by reducing friction.
(3) energy causes friction.
(4) friction causes energy.
(5) energy increases as friction increases.

3. Based on your preview, which of the following do you notice about the form of the information provided?
   (1) There is a graph.
   (2) There is a column of numbers.

(3) There is a diagram showing a series of steps.
(4) The information is provided briefly at the beginning of the item.
(5) There is an article followed by two items.

4. Take the incomplete statement that begins the item ("Since energy is lost when there is friction") and rewrite it as a question. Explain the benefit you get from rewording the information.

---

5. Suppose you were presented with a graph captioned, "Typical Changes in Person B's Body Temperature Over a 24-Hour Period." During your preview, you would probably notice that one side of the graph is labeled: "A.M.—P.M." and the other side is labeled __________. A question you might ask yourself about the graph is: __________.

Items 6 to 9 refer to the following passage.

Living things are grouped into two categories: plants and animals. The green substance chlorophyll is contained in many plant cells and a few animals cells. Which of the following is true?

(1) A cell without chlorophyll must be an animal cell.
(2) A cell with chlorophyll is more likely to be a plant cell than an animal cell.
(3) A cell with chlorophyll is more likely to be an animal cell than a plant cell.
(4) Human cells contain chlorophyll.
(5) Cells from a lettuce leaf do not contain chlorophyll.

6. In your preview, you should have noticed a clue about how many items are based on the information about plant and animal cells. Identify the clue words and tell why they "jump off the page."

7. The questions you keep in mind as you read the short passage include those you make up as well as those that appear ______.

8. Make up a question based on the first sentence of the passage. Explain whether it is answered in the second sentence.

9. Explain the following statement: It doesn't really matter whether the question you asked in Item 8 is ever answered.

---

10. Number the following steps from 1 to 5 to indicate the order in which they should be taken.

a. Predict answers to the questions, including test item(s).
b. Glance at the form of the material and its item(s).
c. Examine each of the five choices you are given.
d. Ask questions while reading the material.
e. Decide on the correct answer.

Answers are on page 253.

Chapter

2

Comprehension

Objective

In this chapter, you will learn about three kinds of comprehension items that you will meet on the GED science test. You will read about and practice

- Restating information
- Summarizing information
- Identifying implications (underlying meanings)

Lesson 1 Restatement

Why is it useful to be able to restate an idea?

When you restate an idea, you put it in your own words or you identify it when you see it in another form. Suppose your friend rushes up to you and says, "Rosemary has lost her marbles!" You reply, "You mean she's gone crazy?" You are testing your comprehension of your friend's statement by putting it into your own words.

Restating Information in Paragraphs

Read the following paragraph and answer the question.

Cities, towns, and many country houses have systems that supply water under pressure. Water pressure can be created in several ways, but the easiest way is by using gravity. Gravity, of course, is the force that pulls all things on earth downward. By building a reservoir on a mountain, engineers can use gravity to pull the water downward into pipes leading to the city.

Which phrase best restates the definition of gravity in the passage?

(1) the easiest way to maintain water pressure
(2) the only way to maintain water pressure
(3) seriousness, as of a situation
(4) a power that pulls everything toward the earth
(5) the force that can pull water into pipes

The correct answer is **(4)**, which restates the definition in the third sentence of the passage. Alternatives **(1)** and **(5)** are also true, but they are not good definitions of gravity, because they are too specific. Alternative **(2)** is contradicted by the second sentence in the passage. Alternative **(3)** is a definition of another meaning of gravity.

Restating Information from Graphics

Study the graph in Figure 1. (You may wish to put a marker on this page, since this diagram is referred to several times in the unit.)

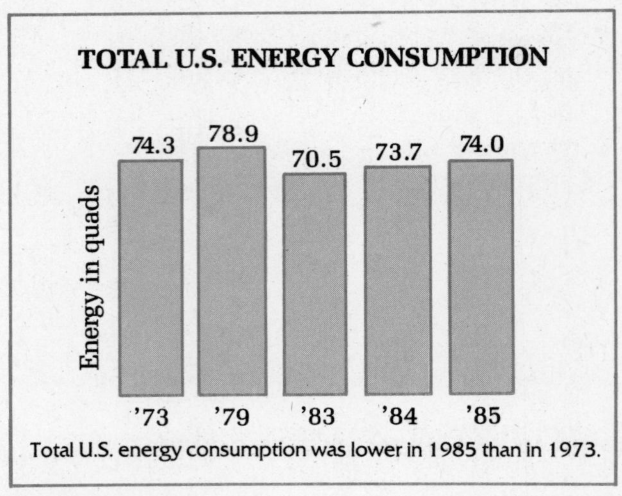

Total U.S. energy consumption was lower in 1985 than in 1973.

FIGURE 1

Note that the caption uses words to restate some of the information given in the graph. The main idea in the caption is that the United States used slightly less energy in 1985 than in 1973.

Tips on Restating the Main Idea

1. Use clues to identify the topic. In paragraphs, repeated words are clues; in graphics, titles and captions are clues.
2. Find the most important thing said about the topic.
3. Restate that information in your own words.
4. In a multiple-choice question, choose the best restatement from the five alternatives.

Lesson 1 Exercise

Items 1 and 2 are based on the following paragraph.

A serious germ-borne disease is usually divided into three stages. The first is known as the *incubation* period. During this stage, germs enter a part of the body where they begin to grow and reproduce. The second stage, the *acute* period, occurs when the symptoms of the disease appear and the person feels sick. This is a sign that the body's defenses are fighting the germs. If and when the germs are destroyed, the patient enters the period of *recovery*. The body functions return to normal as the person gets better.

1. Which sentence best states the main idea?
   (1) The length of the incubation period varies with the disease.
   (2) The patient feels ill during the acute period.
   (3) The acute state is usually the longest of the three periods.
   (4) Severe diseases caused by germs often occur in three steps.
   (5) The stage of recovery is the shortest.

2. Restate the definition of the incubation period in your own words.

Item 3 is based on Figure 2. You may wish to put a marker on this page, since this graph is referred to several times in the unit.

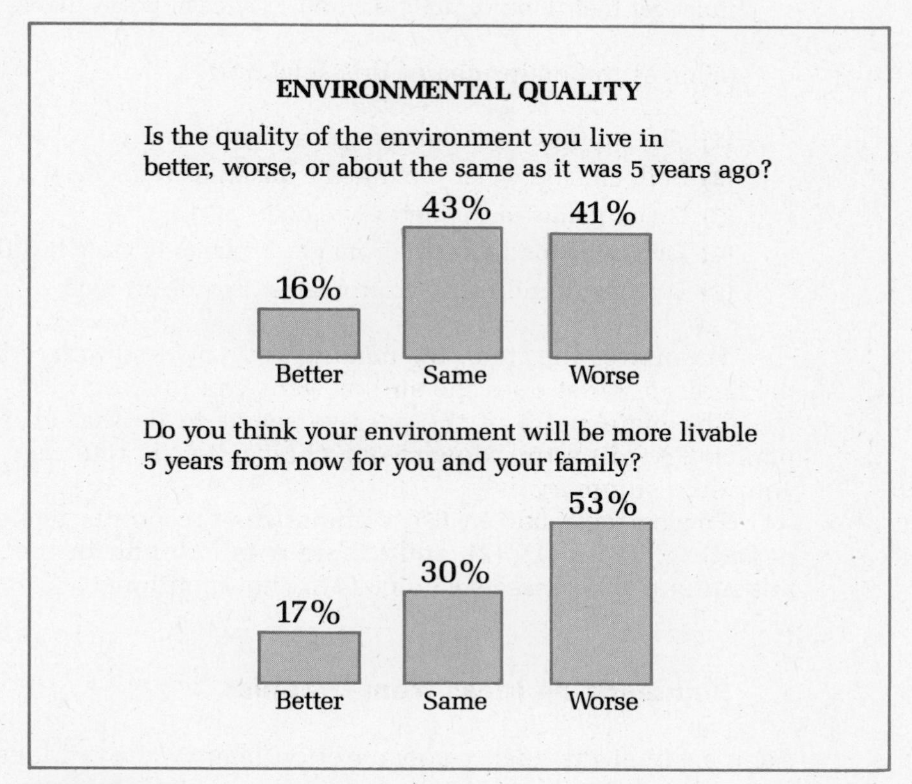

FIGURE 2

3. Restate the information in the graph to describe how most people who answered the poll feel their environment will change over the next five years.

Answers are on page 254.

Lesson 2 Summarizing Ideas

How do you summarize the ideas found in a passage or graphic?

When you summarize ideas, you list the most important ones. After you have seen a new movie, you may tell a friend what it was about. As you condense a couple of hours of action into a few minutes of description, you are summarizing what you saw on the screen.

Summarizing Information in Paragraphs

Read the following paragraph and answer the question.

As you already know, all living things must get food for energy and materials. The energy in food really comes from the sun, while the materials in food come from soil, air, and water. Green plants use sunlight to change carbon dioxide, water, and minerals into food. Almost all other living things must get their food directly or indirectly from green plants.

What is the main idea of this selection?

(1) The sun produces the energy in food.
(2) Soil, air, and water provide the materials in food.
(3) Green plants make their own food.
(4) Obtaining food directly from green plants is more healthy.
(5) Directly or indirectly green plants provide all food.

To answer the item, try making your own summary before looking at the choices. What does the author want you to know?

The main point of the passage seems to be that all food comes from plants. Try to narrow down your choices to the one that most closely fits your own summary.

The correct answer, **(5)**, summarizes the points made throughout the passage. Choices **(1)**, **(2)**, and **(3)** are details found in the passage; they do not sum up the passage. Choice **(4)** is not mentioned.

Summarizing Ideas from Graphics

Look again at the energy-consumption chart you examined in the previous lesson, page 32. After 1973, energy consumption in the United States

(1) rose, then dropped and started rising again
(2) fell, then rose and started dropping again
(3) rose every other year
(4) rose steadily to an all-time high in 1985
(5) dropped steadily to an all-time low in 1985

When previewing the graph, you should have noticed that only some of the years following 1973 are shown. You do not know how much energy was used every year, so eliminate Choice **(3)**. You can summarize the trend, or general change, by simply comparing the heights of the bars. After 1973, the first bar rises, the next bar drops, and the next two rise. Choice **(1)** correctly describes this trend.

Tips on Summarizing

1. To summarize the ideas in a passage, ask yourself: What is the important point to which each of these sentences is building? The summary may or may not be provided directly in the passage.
2. To summarize the information in a graphic, ask yourself: What does the overall picture look like? Pay attention to trends such as those shown by the direction of a line in a line graph or bars in a bar graph.

Lesson 2 Exercise

Items 1 to 3 are based on the following paragraph.

"It's an old story that running wears out joints," says Dr. Nancy Lane, who directed a Stanford study comparing the joints of 41 long-distance runners with those of 41 nonrunners and occasional runners. The researchers found no difference between the two groups in the prevalence of osteoarthritis, or degenerative joint disease. However, the runners ages 50 to 72 did have 40% higher bone density than their counterparts in the control group. "Running prevents bone loss," concludes Lane, "and that's a good finding for women," since they often develop osteoporosis after menopause.

1. TRUE or FALSE: According to Lane's study, running wears out joints. Explain your answer.

2. List two ways in which the two groups in the study are different. List two ways in which the two groups are alike.

3. Lane's study

(1) confirms a popular belief about harm caused by running
(2) contradicts a popular belief about harm caused by running
(3) indicates that running is good for the heart
(4) indicates that running decreases risk of osteoarthritis
(5) reveals that running decreases bone density

Answers are on page 254.

Lesson 3 Identifying Implications

How do you identify the implications in a passage or graphic?

People sometimes imply something without coming right out and saying it. Suppose your employer shakes her head and comments, "It's ten o'clock" when you arrive at work late. You can infer, or figure out, the underlying message: You should have arrived an hour earlier.

Identifying Implications in Paragraphs

Reread the passage about running on page 35. Does the author imply that having higher bone density is desirable?

You are told that runners have higher bone density than nonrunners. You learn that the prevention of bone loss by running is "a good finding for women," since women often develop osteoporosis. Read between the lines. You can infer that having higher bone density (heavier, stronger bones) is a good thing because higher bone density helps offset osteoporosis.

Identifying Implications in Graphics

Look again at the diagram about energy consumption on page 32. What is implied about energy consumption between 1973 and 1979? There are no bars shown for these years, but you can do some mental detective work to figure out what the bars would show. Since 74.3 quads were consumed in 1972 and 78.9 quads were used in 1979, the figures for 1973–79 probably lie somewhere between 74.3 and 78.9.

Tips on Identifying Implications

1. Ask yourself what facts the author has provided.
2. Examine the relevant clues among these facts.
3. "Read between the lines" by connecting the clues through logic.

Lesson 3 Exercise

Items 1 to 3 are based on the following paragraph.

A tick about the size of the head of a pin is causing a lot of trouble, especially in the Northeast. The tick causes an illness in humans called Lyme disease. Since its identification, the disease has caused only one human death, but it can trigger chills, fever, fatigue, and arthritis. Some residents of the Northeast are asking permission to kill deer because the tick needs them in order to reproduce.

1. TRUE or FALSE: The tick that causes Lyme disease is difficult to see. Explain your answer.
2. People who get Lyme disease are certain/likely/unlikely to die from it. (Choose one.)
3. What can you infer about deer?
   (1) More deer would result in less Lyme disease.
   (2) Fewer deer might result in less Lyme disease.
   (3) More deer might result in fewer ticks.
   (4) More ticks would result in fewer deer.
   (5) Fewer ticks would result in more deer.

Answers are on page 254.

Chapter 2 Quiz

1. Suppose you come to a diagram on the Science section of the GED with the caption "Two-Year Life Cycle of the Tick." Briefly explain how you would apply what you have learned about previewing and questioning.
2. Match each of the three types of comprehension items on the left with the appropriate description on the right.

(1) restating information	a. condensing the important points
(2) summarizing ideas	b. "reading between the lines"
(3) identifying implications	c. putting into other words

3. Label the following three items according to what each requires the reader to do. R = restate information; S = summarize ideas; I = identify implications.
   (1) What can be inferred about continued use of DDT?
   (2) According to the first line, why is DDT used in Third World nations?
   (3) Which would be the best title for the article about DDT?

Items 4 to 7 are based on the following passage.

Water buffalo could become the American farm animal of the future. Blind taste tests show that most people find water buffalo meat tastes at least as good as beef and is just as tender. In addition, laboratory analysis shows it has less fat and much less cholesterol. Water buffalo milk has twice as much butterfat as cow's milk and 50 percent more protein. Best of all, water buffalo have a more efficient digestive system that makes them grow sleek and fat on poor, scrubby, or swampy rangeland that barely keeps cattle skinny.

4. Three advantages to using water buffalo instead of cattle are: ______.

5. Briefly explain in your own words what the phrase *swampy rangeland that barely keeps cattle skinny* means.

6. TRUE or FALSE: By comparing cholesterol, fat, and protein in water buffalo meat with that in beef, the writer is implying that eating water buffalo meat is more healthful. Explain your answer.

7. Locate the sentence from which it can be inferred that raising water buffalo may be easier than raising cattle.

Items 8 to 10 are based on the following drawing.

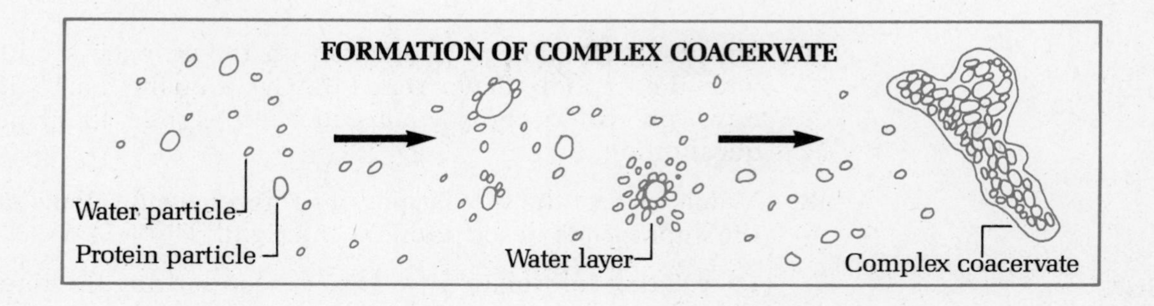

This drawing shows how a complex cluster of proteins is formed when a water layer surrounds a group of protein molecules.

8. Briefly describe in your own words the two steps shown in the diagram.

9. As shown by the diagram, there are more ____________ particles than ____________ particles.

10. Which of the following is NOT implied by the diagram?

(1) Something attracts water particles to protein particles.
(2) Water makes up the outer layer of the protein clusters.
(3) Water particles are larger than protein particles.
(4) Some water particles do not surround protein particles.
(5) Most of the protein particles do become surrounded by water.

Answers are on page 254.

Chapter

3 Application

Objective

In this chapter, you will learn about a second level of item on the Science section of the GED, the *application* items. You will read about and practice:

- using ideas in a situation other than that provided in a passage or graphic
- categorizing ideas according to given definitions

Lesson 1 Application from Passages and Graphics

How do application items differ from comprehension items?

To apply something means to use it. Suppose the weather forecast says, "Snow will blanket the entire eastern half of the country tomorrow." You live in Boston, so you pull out your boots. You are applying what you know to the predicted situation. Application items on the Science section of the GED test your ability to take information you are given and use it in a new situation. Application is a higher level thinking skill than comprehension. Comprehension requires understanding of information, whereas application requires using this understanding to make a prediction.

Application of Information from Passages

Read the following paragraph and answer the question.

At Chicken Soup, a day-care center for sick children, the Polka Dot Room handles children with chicken pox. Those with tummy aches go to the PopSicle Room, and kids with colds are assigned to the Sniffles Room. The Minneapolis center is one of a growing number of sick-care services being developed nationwide to provide working parents with alternatives to absenteeism when a child becomes ill.

Chicken Soup would be most appropriate for which of the following?

(1) Joslin, a 3-year-old with severe pneumonia
(2) Jake, a 4-year-old with severe behavior problems
(3) Elvia, a 2-year-old with a stomach ache
(4) Adam, a 4-year-old in perfect health
(5) Angela, a 16-year-old who has the measles

The correct answer is **(3)** because Elvia is a child with a minor physical complaint. You should eliminate **(1)** because Joslin's illness sounds more serious than the usual cases of chicken pox, stomach ache, and cold that the center handles. Eliminate **(2)** for a similar reason; the center is for children with childhood illnesses, not behavior problems. Choice **(4)** describes a child who belongs in a regular center, since he is not sick. Angela, in Choice **(5)**, is too old for day care.

Notice how important it is to comprehend what the purpose of the center is before you can apply the information in the passage to specific new cases. Notice, too, that you cannot make a prediction about the answer before you look at the alternatives, because each alternative contains information that is not in the passage.

Applying Information from Graphics

Look again at the diagram about environmental quality on page 33. A woman has seen the woods where she played as a child cut down to put up a housing development. The creek where she used to swim is polluted now, and the air is thick with smog from new factories. If the woman were one of the 1500 readers polled, do you think she would be in the group of 17 percent, 30 percent, or 53 percent?

To answer this question, you need to make sure you understand both the "old" and "new" situations. Then build a bridge of ideas between them. The illustration shows that 17 percent of the people feel the environment will become better, 30 percent feel it will remain the same, and 53 percent fear it will become worse. The "new" information in the item tells you that the woman has seen her environment become worse. By putting the clues together, you can deduce that the woman would be among the 53 percent who fear the environment will grow worse.

A Strategy for Answering Application Questions

1. Summarize for yourself both "old" and "new" information.
2. Figure out what ideas are common to both.
3. Examine all test alternatives carefully.
4. Check each choice against the original passage or graphic by asking: Does this new information fit with what I already know?

Lesson 1 Exercise

1. TRUE or FALSE: Although you can find the answer to application items in the material, you read for information to use in arriving at the answer to comprehension items. Explain your answer.

2. Identify the type of item presented by labeling each: C = Comprehension; A = Application.

   a. According to the diagram, what causes lightning?
   b. Based on the description of sedimentary rock, would you say that such rock is more like a club sandwich or a milkshake?
   c. In a chemical solution, a gas, a liquid, or a solid is mixed in a gas, a liquid, or a solid without a chemical change. Which of the following household products is an example of a solution?
   d. What explanation does the paragraph give for why it is easier to move a refrigerator if you have a dolly (moving cart)?

Item 3 is based on the following passage.

In its last major survey of erosion four years ago, the Department of Agriculture estimated that roughly three billion tons of soil were being stripped from the nation's farmland each year. Now, many experts believe, the problem may be even worse. This is due in part to the fact that farmers can often get government money if they cultivate fragile land. This land is easily eroded, or worn away by wind and water. Already such erosion has reduced many farmers' crop yields. Eventually the nation's food-producing ability may be endangered.

3. According to the passage, farming fragile land is most similar to winning

   (1) a scholarship that enables you to get the degree you want
   (2) a wood stove that starts a destructive fire in your home
   (3) a free meal at a restaurant you usually cannot afford
   (4) a trip to a region you have always wanted to see
   (5) an air conditioner that you don't really need

Answers are on page 254.

Lesson 2 Application from a List of Categories

What does the "new item type" look like?

A birdwatcher spots an unusual bird and tries to figure out what type of bird it is, based on what he or she knows about various categories of birds. Similarly, on the GED you will be presented with items that require you to apply what you learn from a list of definitions. Items in this new format require that you first understand the definitions presented and then match specific cases with the proper definition.

The following is an example of application items in the new format. Examine the definitions and answer the following questions.

Below are descriptions of five parts of the human brain:

(1) **cerebellum**—controls posture and coordination
(2) **hypothalamus**—controls blood pressure and appetite
(3) **pineal gland**—acts as an internal clock
(4) **cerebrum**—responsible for most thinking and learning
(5) **limbic system**—responsible for triggering emotions

1. Nikki experiences a surge of joy as she crosses the finishing line. Which part of her brain is primarily responsible for her feeling?

(1) cerebellum
(2) hypothalamus
(3) pineal gland
(4) cerebrum
(5) limbic system

2. Sam delicately threads a needle. Which part of his brain is primarily responsible?

(1) cerebellum
(2) hypothalamus
(3) pineal gland
(4) cerebrum
(5) limbic system

During your preview, you probably noticed that the choices in both items are the same. That is because both items are testing your understanding of the different roles played by the various regions of the brain. Notice, also, that you are not expected to "pull information out of the air" about these regions. Success on the GED depends on careful reading of what is provided, rather than on remembering isolated facts you have previously learned.

In the first item, the correct answer is **(5).** Joy is an emotion. As you review the list of descriptions, you see that it is the limbic system that controls emotion.

The correct choice for the second item is **(1),** since the cerebellum controls coordination. You need to be able to coordinate your hand with your eye to thread a needle.

A Strategy for Answering "New Item" Format Questions

It is not necessary to take the time to memorize what each part of the brain does. A more efficient strategy is to read the "new" information in the item and compare it against each of the pieces of "old" information in the list. For example, you could ask yourself: Does threading a needle involve coordination **(1)**? Yes. Then consider the other choices to make sure that there is not a better one. Does successfully threading a needle depend mainly on blood pressure? Temperature? Not really. Thinking? Maybe. Inner clock? Emotion? No. You have narrowed your choices down to **(1)** and **(4)**. The best answer still seems to be **(1)**.

Lesson 2 Exercise

Items 1 to 3 are based on the following information.

Living cells often have a greater chance of surviving when they cooperate with each other instead of competing. Categories of organization include the following:

(1) **cell part**—each tiny part of a cell has a different role
(2) **cell**—together, the cell parts function as a living unit
(3) **tissue**—similar cells function together, as in muscle tissue
(4) **organ**—different tissues function together, as in a kidney
(5) **organ system**—different organs function together, as in the circulatory system

1. A scientist takes a tiny cutting from a plant and examines it under a microscope. He observes several oval units, each the same as the next. He is probably looking at a(n)

(1) cell part
(2) single cell
(3) tissue
(4) organ
(5) organ system

2. A surgeon transplants a heart from one person into another. The heart, containing muscle and nerve tissue, is an example of a(n)

(1) cell part
(2) single cell
(3) tissue
(4) organ
(5) organ system

3. Several organs work together to convert the apple you eat into the energy you use. The digestive tract responsible for the conversion is a(n)

(1) cell part
(2) single cell
(3) tissue
(4) organ
(5) organ system

Answers are on page 254.

Chapter 3 Quiz

1. Do you think that most application items are easier or more difficult than comprehension questions? Why?

2. TRUE or FALSE: Application items on the Science section of the GED will be based on passages, but not on graphics (tables, graphs, diagrams, and so forth). Explain your answer.

3. Suppose you preview a set of items on the Science section of the GED. You notice that all the items have the same five alternatives. Could two items have the same answer?

Items 4 to 6 are based on the following information.

4. In a parasitic relationship, one living thing lives off a second living thing and does harm to the second organism.

Which of the following is an example of a parasitic relationship?

(1) A butterfly is left alone by birds because it looks like another, bitter-tasting butterfly.
(2) A mold spreads over grape jam and consumes it.
(3) Two plants grow together, each supplying something the other needs to live.
(4) An eel attaches itself to a trout, releasing the trout only after it has killed and eaten it.
(5) A fungus grows on a decaying stump of a dead tree.

5. TRUE or FALSE: The relationship of a tapeworm to a human being is an example of a parasitic relationship.

6. Can you think of an example of two organisms in a parasitic relationship?

Items 7 to 10 are based on the following graph.

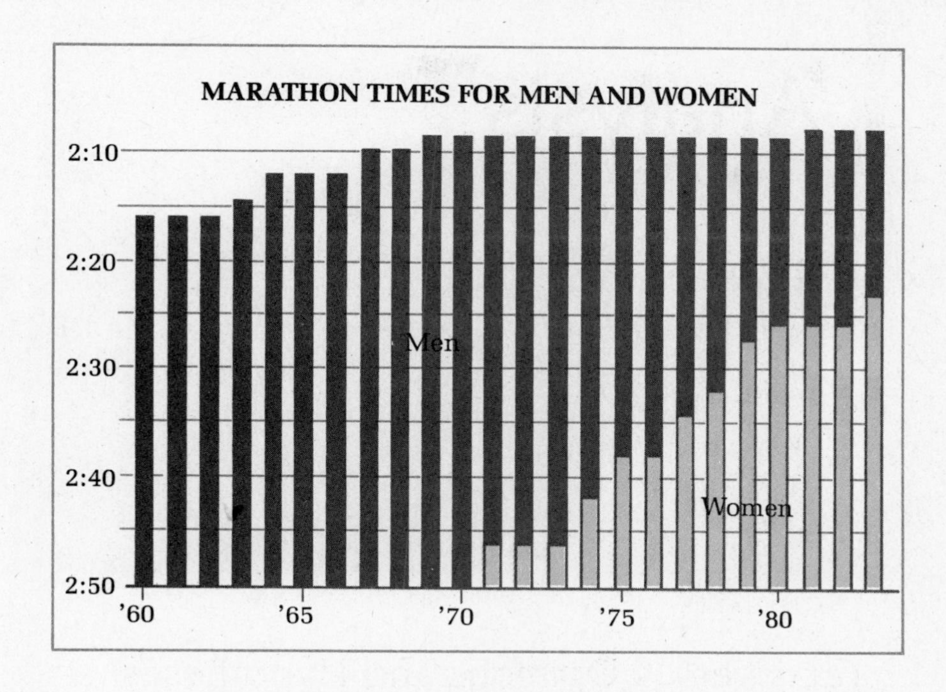

7. The relationship between women's and men's times in the marathon between 1971 and 1984 is most like the relationship between

(1) two fish swimming side by side in a tank
(2) two birds flying in a migratory pattern
(3) two racing cars, the slower gaining on the faster
(4) two bikers going in opposite directions on a hill
(5) two leaves, one drifting downward and the other blown upward

8. If women had started running in marathons in 1960 instead of 1971, how do you think the illustration would be different?

9. Briefly explain another situation in which women have lagged behind men over the years but are now approaching equality.

10. TRUE or FALSE: The information shown in the bar graph could also have been presented in a line graph with one line showing men's changing time over the years and one line showing women's changing time over the years.

Answers are on page 255.

Chapter 4

Analysis

Objective

In this chapter you will learn about a third level of item that you will meet on the GED Science test—the analysis item. You will read about and practice:

- Distinguishing facts from hypotheses or opinions
- Recognizing unstated assumptions
- Distinguishing conclusions from supporting statements
- Identifying cause-and-effect relationships

Lesson 1 Facts versus Opinions and Hypotheses

How can you distinguish facts from opinions or hypotheses?

To analyze something means to look at the relationships between its parts. When laboratory technicians analyze blood samples, they look at the elements that make up the blood. Analysis items on the GED tests assess your ability to understand how separate ideas are organized or relate to each other.

One type of analysis item on the GED asks you to point out which of several statements are facts and which are opinions or hypotheses.

Questions to Help Distinguish Facts, Opinions, and Hypotheses

1. Ask yourself: Is this something that can be proved by measurement or observation? If so, it is a fact.
2. Is this someone's personal interpretation of the way things are, a view that cannot be proved? If so, it is an opinion.
3. Is this an educated guess or theory developed to explain certain facts? If so, it is a hypothesis.

Distinguishing Facts from Hypotheses in a Graphic

Look again at the graph comparing men and women marathon runners on page 45. Which of the following is a hypothesis that could be used to explain the fact that women have cut so much more off their record time than men have during the same ten-year period?

(1) Women began running in world marathons in 1960.
(2) In ten years, women reduced their record time by 20 seconds.
(3) Women are such inferior athletes that they will never equal men.
(4) Because men were initially better trained, they were already running close to their top potential.
(5) Ten years ago, women were running at closer to the top possible speed than men were.

The correct answer is **(4).** You can easily eliminate **(1)** because it is false, according to the graph. You can also eliminate **(2).** Women did shave 20 seconds off their record, but remember that the question asks for a *hypothesis* to explain the facts. Choice **(3)** is an extreme opinion that would be impossible to prove. Once you have narrowed you choices down to **(4)** and **(5),** you can eliminate **(5),** since it is contradicted by the graph.

The facts show that women now run much faster than they did ten years ago; they were far from their greatest potential speed when they first started competing. A good hypothesis to explain why women improved more than men would be that there was more "room for improvement." Early women runners were more poorly trained than men, who had been competing for years.

Summary of Differences between Facts, Opinions, and Hypotheses

A statement is a(n)

- fact if it can be proved true by observation or measurement
- opinion if it is an emotional interpretation that cannot be proved
- hypothesis if it is an educated guess used to explain facts

Lesson 1 Exercise

Items 1 and 2 refer to the following advertisement.

An advertisement for Swiss cheese states:

A. can prevent sugars from forming harmful acids
B. the delicious cheese Americans love
C. provides calcium, important for dental health
D. used by the finest chefs

BUY HEIDE-BRAND SWISS CHEESE

1. Which of the above statements is most likely based on facts rather than on opinion?

(1) A only
(2) B only
(3) A and B only
(4) A and C only
(5) B and D only

2. Which of the above statements could NOT be proved by chemical tests?

(1) A only
(2) B only
(3) A and B only
(4) A and C only
(5) B and D only

---

3. Label the level of each of the following items. C = Comprehension; AP = Application; AN = Analysis

a. What is the central hypothesis of the study?
b. Energy can be neither created nor destroyed. Which of the following examples illustrates this principle?
c. From the map, what is the time difference between Japan and Boston?
d. Which of the statements provided is most likely based on facts rather than on opinion?

Answers are on page 255.

Lesson 2 Identifying Unstated Assumptions

How can you identify an unstated assumption?

An assumption is an idea that you take for granted without proof. If you stop to fill up your car and find a sign on the pump saying, "Out of order," you *assume* that the sign tells the truth and you look for a pump that works.

Authors, too, make assumptions. On the GED, you will be asked to identify assumptions that are unstated. These assumptions may or may not be accurate. It is very important to avoid being swayed by inaccurate assumptions.

Identifying Unstated Assumptions That Are Accurate

Look at the following passage about malnutrition. When the speaker calls malnutrition a challenge, what is he assuming? He is taking for granted that we can and should do something about the problem. This is an assumption with which few would probably disagree.

> On May 6, 1966, United States Secretary of Agriculture Orville L. Freeman discussed the world population and food problem. He said: "In the developing nations, some 171 million children under six years of age, and some 98 million between six and fourteen, suffer seriously from malnutrition. Millions who survive are permanently handicapped. . . . Against this dark background, mankind faces its greatest challenge."

Identifying Unstated Assumptions That May Not Be Accurate

> Twenty-six countries now produce nuclear-generated electricity. Six more plan to do so by 1990. Japan, France, the Soviet Union, and China are among the many nations committed to nuclear electricity as an economical, safe alternative to oil. Not one future nuclear plant has been planned in the United States since 1978, while at least 50 have been ordered in other parts of the world. In a competitive world market where abundant energy is a must, can America afford to fall behind in the very technology we pioneered?

The assumption the writer makes is that we cannot. However, there are those who argue that we can and should cut down on our use of nuclear power. The writer assumes that nuclear electricity is "safe," despite a number of nuclear accidents. The writer also assumes that it is very important that the United States hold a lead in technology and in the world economy. Others might argue that environmental and safety concerns are more important than keeping up front in the world market.

Identifying Assumptions in Graphics

Look again at the chart about environmental quality on page 33. Which of the following is an unstated assumption?

(1) 41% feel the environment is worse than it was 5 years ago.
(2) 30% feel the environment is the same as it was 5 years ago.
(3) The majority feel the environment will be worse in 5 years.
(4) People responded to the poll truthfully.
(5) People polled may or may not be interested in nature.

The correct answer is (4). Whenever you see the results of a poll or questionnaire mailing, the assumption is made that the answers represent what the people really believe. Sometimes, however, people either misunderstand what they are being asked or are influenced somehow to answer inaccurately.

Choices (1) and (3) can be eliminated; they describe facts that are stated in the graphic. Choice (2) is a mistaken reading of the table; 30 percent feel the environment *will be* the same in five years. You can eliminate Choice (5), since a careful reading of the caption shows that those polled were all readers of *National Wildlife*; the assumption is that all such readers do have an interest in nature.

How To Identify an Unstated Assumption

1. Don't accept every statement at face value. Ask yourself: What is the unstated, underlying belief that I am being asked to accept?
2. Make sure that the statement you think to be an unstated assumption is not a stated fact, opinion, or hypothesis.

Lesson 2 Exercise

Items 1 to 3 are based on the following passage.

Underwater junkyards are a key part of a nationwide effort to create artificial reefs that will serve as breeding habitats for fish. The reason: Ninety percent of the ocean floor off U.S. coasts is a sand and mud wasteland. Artificial reefs are successful in hiding small fish from predators. They also reduce crowding on natural formations by offering growing surfaces for plants and more food.

1. What unstated assumption does the author make?

(1) Humans should help create artificial breeding grounds for fish.
(2) Small fish use underwater junkyards to hide from predators.
(3) Oil spills have destroyed natural reefs.
(4) Mud and sand occupy most of the ocean floor.
(5) We should not let our junk get into the ocean.

2. TRUE OR FALSE: The writer is assuming that the benefits of having underwater junkyards outweigh any problems such junkyards may create. Explain your answer.

3. Suppose an ecologist writes a reply to the passage about artificial reefs. The ecologist is worried about the possibility of upsetting the balance of nature. Complete this reply: You praise the reefs for hiding small fish from predators. Are you sure you're right to assume that it is more important for small fish to breed than it is for predators to __________?

Answers are on page 255.

Lesson 3 Conclusions versus Supporting Statements

How can you distinguish between a conclusion and supporting statements?

A conclusion is a final statement summarizing your main idea and the decision or opinion that you have reached. Your conclusion is often based on several supporting details or facts that point in a particular direction. If a friend is late every time you agree to meet, you may reach the conclusion that he or she doesn't try very hard to be on time.

Writers use supporting details to draw conclusions. On the GED, you will be asked to distinguish between the supporting details an author uses to develop an idea and the final conclusion that is drawn. Ask yourself: Is this statement just one of the building blocks on which the main "room" rests, or is it actually the writer's main reason for presenting the information?

Distinguishing Conclusions from Supporting Details in a Passage

Reread the advertisement for Swiss cheese on page 48. Which of the following is the conclusion drawn by the advertiser?

(1) Swiss cheese keeps harmful acids from forming.
(2) Americans love Swiss cheese.
(3) Swiss cheese is a source of calcium.
(4) Good cooks use Swiss cheese.
(5) You should buy Heide brand Swiss cheese.

You can eliminate Choices **(1)**, **(2)**, **(3)**, and **(4)** because they are all supporting details the advertiser uses to build up to his main conclusion: You should buy Heide-Brand Swiss cheese **(5)**. Notice that here, as in many cases, the conclusion is at the end of the passage. Notice, also, that some of the supporting details are facts and others are opinions.

Distinguishing Conclusions from Supporting Details in a Graphic

Look again at the graph about energy used in the United States, page 32. Which of the following is the author's conclusion?

(1) 74.3 quads of energy were used in 1973.
(2) Less energy was used in 1983 than 10 years earlier.
(3) 73.7 quads of energy were used in 1984.
(4) There has been a recent increase in energy use.
(5) Americans were using more energy in 1973 than they used in 1985.

Choice **(5)** is the *conclusion* drawn by the writer. You are not asked to identify facts used to support the conclusion—Choice **(2)**. Choice **(3)** describes a misreading of the graph. There has been a recent increase in energy use. However, the key phrase in the caption, "Despite a recent increase . . ." tells you that the important message is that, "energy consumption is still lower today than in 1973." Read captions and headings of graphics carefully when trying to determine the conclusion to be drawn from them.

Remember, supporting details are the pieces of evidence the writer uses to support his main idea. The conclusion is the main idea toward which the writer builds her or his case.

Lesson 3 Exercise

Items 1 to 3 are based on the following passage.

Biochemists are working on a simple blood test to determine if a person has cancer. They have isolated a protein that seems to exist only in the blood of someone with an active tumor. Researchers looked for the marker protein in the blood plasma of 300 patients. They found it only in plasma from 100 patients known to have cancer. In one case, the team detected the protein in a patient thought to be healthy, and then a conventional exam revealed a cancerous tumor. In another case the test showed no sign of the protein in a patient with a tumor, and doctors discovered that the tumor was not cancerous.

1. Which statement best states the conclusion of the author?

(1) Cancer-causing proteins are found in cancer victims' blood.
(2) Marker proteins in blood can be used to identify cancer victims.
(3) A "healthy" person with the marker protein actually had cancer.
(4) If Americans consumed less protein they would have less cancer.
(5) Cancer is one of the leading causes of death in America.

2. Briefly list two of the supporting details the author uses to support his conclusion.

3. Suppose the author wants to add another supporting detail, so he describes the results of a second study done on 200 patients known to have cancer and 100 thought to have other illnesses. Write one sentence describing the results of the study.

Answers are on page 255.

Lesson 4 Identifying Cause-and-Effect Relationships

How can you tell which is the most likely explanation for a change?

In daily life, we often try to find reasons or causes for the changes around us. If your dog starts barking, you usually try to see what is causing the uproar. We also try to predict the effects of many actions. When you are applying for a job, you probably think about how your earnings and quality of life would be affected by your getting the job.

One important way that authors develop their ideas is by showing how one thing causes or results from another. You can identify causes asking yourself: What could be behind this change? You can identify effects by asking yourself: What changes would result from that action? Use the facts presented, background knowledge you already have, common sense, and imagination to make cause-effect connections.

Identifying Cause-and-Effect Relationships in Passages

Look again at the passage about underwater junkyards, page 50. Which of the following statements gives the main reason for creating the junkyards described in the passage?

(1) easy disposal of junk
(2) part of a nationwide effort to build underwater junkyards
(3) concern for clean water
(4) provision of a safe breeding ground for small fish
(5) creation of a trap for the predators of small fish

You need to read the question carefully. You are to give the *main reason* for the creation of the junkyards. Choice (1) may be one of the reasons, but it is not mentioned in the passage. Choice (2) is stated in the passage, but it does not explain the *cause* behind the nationwide effort. Eliminate Choice (3). No connection is made between clean water and junkyards; if anything, clean water advocates would *not* want the junkyards.

You have narrowed your choices down to (4) and (5). Choice (5) is not mentioned in the passage, but this is not enough reason to eliminate it. Sometimes you need to supply a likely explanation where none is given. In this case, however, the explanation for the junkyards, creation of safe breeding grounds (4), is given.

Remember, when trying to establish cause-and-effect relationships, ask yourself what happened and why it happened.

Lesson 4 Exercise

Items 1 to 3 are based on the diagram on energy use on page 32.

1. Which of the following is most likely to explain why Americans used less energy in 1985 than in 1973?
   (1) Americans used energy more efficiently.
   (2) Americans paid little attention to conserving energy.
   (3) Americans used more energy in 1973 than they did in 1985.
   (4) The population of the United States rose steadily between 1973 and 1985.
   (5) More Americans are going to college than ever before.

2. What do you think is one result of the fact that Americans used less energy in 1985 than they used in 1973?

3. Complete the following sentence: One reason that Erin Lux used less energy in 1985 than she did in 1973 is that she ______.

Answers are on page 255.

Chapter 4 Quiz

1. Label the following items according to whether they mainly
   a. test your understanding
   b. test your ability to apply your understanding to a new situation
   c. test your ability to examine the relationships between ideas
   (1) What is the most likely explanation for the change shown?
   (2) If the temperature of gas in a container rises, the pressure increases. What happens to a tire on a hot day?
   (3) According to the passage, what is inertia?

Items 2 to 4 are based on the following passage.

(1) A healthy mouth and teeth communicate more than just words. (2) Your smile can make a business statement appear as elegant as a strand of pearls. (3) It can also project power, self-assurance, and charisma. (4) It makes good business sense to invest some time in daily preventive care and in regular dental visits.

2. Restate sentence (3) in your own words.

3. Label each sentence in the passage F (fact) or O (opinion).

4. Which sentence contains the conclusion?

Items 5 to 6 are based on the following figure.

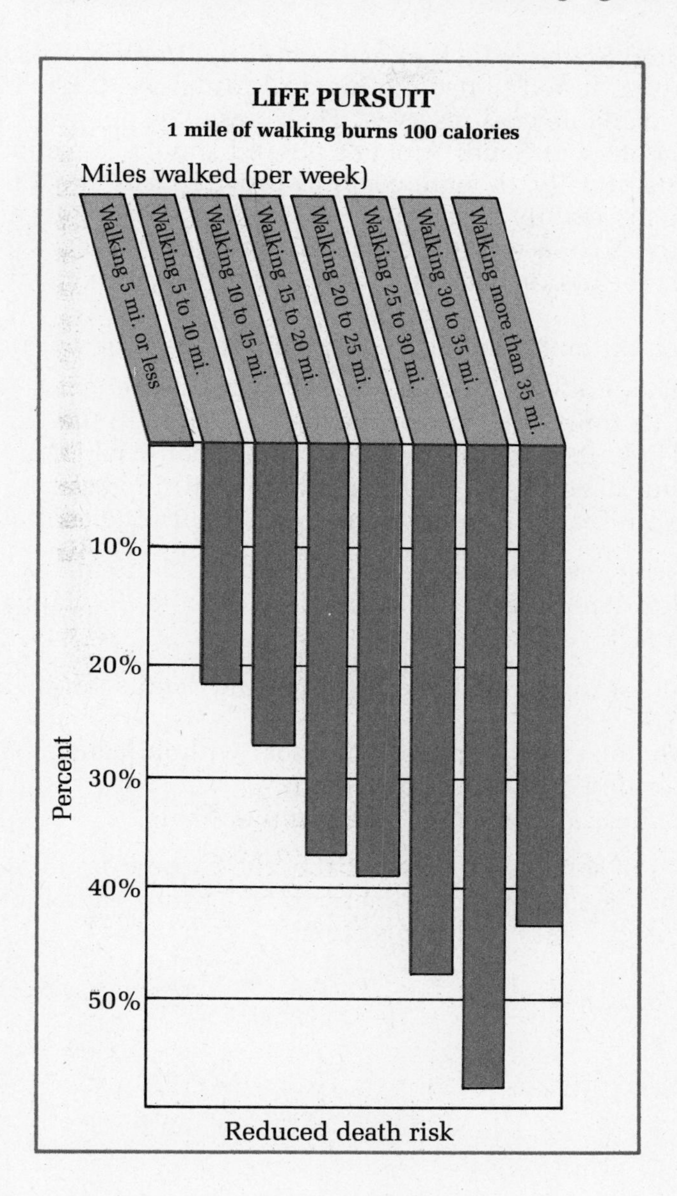

5. According to the graph, which of the following is NOT an effect of walking 39 miles a week?
   (1) It is likely to increase your life span.
   (2) It burns up 3000 calories per week.
   (3) It is a less effective way of reducing death risk than walking 30–35 miles per week.
   (4) It is a more effective way of reducing death risk than walking 20–25 miles per week.
   (5) It reduces death risk by about half.

6. What conclusion do you think the person who developed the graph wants you to draw?

Items 7 to 10 are based on the following passage.

The United States, which already gets about half its electricity from coal, is well suited to developing coal resources. It has 25 percent of the world's available coal reserves. These reserves possess six times the energy content of Saudi Arabia's oil and should last another 200 to 300 years. Historically, though, coal has proved itself far from perfect as a fuel. This is mainly because it releases such pollutants as sulfur dioxide and nitrous oxide when it burns. At the least, these byproducts create the basis for photochemical smog. Worse, they can cause acid rain.

7. Which of the following does the author assume but not state?

(1) The problems of coal use outweigh the benefits.
(2) One-fourth of the world's usable coal is in the United States.
(3) There is more oil in Saudi Arabia than coal in the United States.
(4) Our need for an alternative to oil will increase.
(5) No research is being done to make burning coal cleaner.

8. TRUE OR FALSE: One fact stated in the article is that when burned, coal produces pollutants such as sulfur dioxide. Explain why this is or is not a fact.

9. Which of the following are supporting details, and which is a conclusion?

a. We must find a way to burn coal without causing acid rain.
b. Burning coal causes acid rain.
c. Acid rain kills many plants and animals.

10. Read the statements below. Label the cause C and the effect E.

a. Coal is considered by many to be an imperfect fuel.
b. Coal releases pollutants.

Answers are on pages 255–256.

Chapter

5 Evaluation

Objective

In this chapter you will learn about a fourth level of item on the GED science test, the evaluation item. You will read about and practice:

- assessing whether a conclusion can properly be drawn from data
- recognizing the role of values in decision making
- assessing whether facts are accurate
- indicating logical fallacies in arguments

Lesson 1 Adequacy and Appropriateness of Information

How can you judge whether a statement is supported adequately by the evidence presented?

To evaluate something means to judge it. When a teacher gives you a final grade, the teacher is evaluating, or judging, what grade best represents your effort and ability, based on observations of what you have done.

Evaluation items on the GED exam test your ability to judge whether information or methods are accurate. You are often asked to use a standard, or "mental yardstick," provided in a passage or graphic to judge various alternatives.

For example, after examining the graph about walking, page 55, you might be asked to determine which of several statements is best supported by the data (information) in the graph. In each case, you should ask yourself: "Can I find a logical connection between this statement and what I have learned from the graph about the benefits of walking?"

Evaluating Adequacy of Information in a Passage

Read the following passage and answer the question that follows.

> One of the earliest scientific attempts to measure the benefits of eating breakfast was a series of experiments conducted in the 1940s and 1950s by nutrition researchers W. W. Tuttle, Kate Daum, and colleagues at Iowa State University. One area that interested the Iowa researchers was the effect of breakfast on schoolchildren. In a study of seven boys between the ages 12 and 14, the researchers found that boys could not pedal as hard or as long on a bicycle when they missed breakfast.

Which of the following statements is best supported by the evidence?

(1) Missing breakfast decreases boys' energy more than girls'.
(2) Boys who miss breakfast will not be able to run as fast.
(3) Men who compete in bicycle races should eat breakfast.
(4) Eating breakfast improves teenage boys' classroom achievement.
(5) Missing breakfast reduces young teenage boys' physical energy.

The correct answer is (5). Since no girls or adult males were included in the study, there is *not enough* information to support Choices (1) or (3). Since running and classroom achievement were not studied, the information is of the *wrong kind* to support Choices (2) and (4). Notice how important it is to read the question and each choice carefully. Some of the choices may well be true statements, but they are not adequately supported.

Key Questions to Ask About Adequacy and Appropriateness of Data

When trying to assess the adequacy or appropriateness of data to support hypotheses, conclusions, or generalizations, ask yourself:

1. Is there enough information to support this statement?
2. Is the information the right kind to support this statement?

Lesson 1 Exercise

1. Read the following statement and indicate whether the conclusion is supported by the information. Explain why or why not.

   Most major scientific discoveries have occurred as a result of painstaking laboratory experiments over a number of years. Therefore, if I spend a long time in the laboratory, I will eventually come up with a major scientific discovery.

2. Suppose a scientist notices that a teaspoon of salt dissolves in a cup of water. Describe a generalization about salt and water that would NOT be adequately supported by this observation.

3. Look again at the graph about marathon runners on page 45. Which of the following statements made in the caption is best supported by evidence presented in the graph?

   (1) Women got off to a slow start.
   (2) Women are catching up quickly.
   (3) Men have struggled to improve their time.
   (4) Men improved their time by a merely minor amount.
   (5) Women improved their time by nearly 20 minutes.

Answers are on page 256.

Lesson 2 Effect of Values on Beliefs and Decision Making

Why is it important to see how values affect a writer's beliefs?

Each of us has a set of values, or ideals and customs, that we consider important. What we think and do is colored by those values. Someone who values scenic beauty is more likely to go camping in the wilderness than someone who does not value scenic beauty.

What authors write is also colored by their values. On the GED, you will be asked to examine the values shaping written material. Once you are aware of these values, you are in a better position to judge whether you want to accept what you are being told. Ask yourself: How do the writer's values affect the way information is presented? Is the author being objective?

For example, reread the passage about eating breakfast on page 57. Suppose the article had ended with the statement "It was the opinion of the school principal and of the teacher that breakfast was a material asset to the boys in both attitude and scholastic accomplishment," Tuttle and Daum wrote. Would you accept the principal's and teacher's idea without questioning it? Preferably not. The principal and teacher probably value doing well in school. They would like to think that student *mental* performance could be improved by eating breakfast, but the study measured only *physical* performance.

Identifying the Effect of Values on Information in Graphics

Look again at the table about environmental quality on page 33. Suppose that the readers polled had all been readers of *Consumer's Magazine*. Do you think the results would have been the same? Perhaps not. Someone who values easy access to shopping malls and consumer services might be more likely to predict that the environment will be "more livable" in five years than someone who values preserving national wildlife. Be careful *not* to assume from the graph that 53 percent of *all* citizens think the environment will be less livable in five years.

Key Questions That Focus on the Effect of Values

To recognize the role values play in beliefs and decisions, ask:

1. What is the source I am reading?
2. What would the writer like me to believe?
3. Has the writer used more than emotional appeal to support this belief?

Lesson 2 Exercise

Items 1 to 3 are based on the following passage.

Some earth scientists think that Alaska's coastal plain could lie over a huge oil field. The nation's oil companies are eager to test this theory by drilling. Opponents of the drilling claim that the area is an important home to as many as 200,000 caribou. They want to add the coastal plain to a 19 million-acre wildlife refuge recently established by Congress.

1. Which of the following statements would be LEAST likely to appear in a promotional ad paid for by oil drillers?

(1) Alaska has recently lost its number-1 position in earnings per person.
(2) The United States should further develop energy sources within the country.
(3) Independence from foreign oil producers is of utmost importance.
(4) Caribou have been grazing contentedly where drilling will occur.
(5) Studies haven't been done on effects of drilling on breeding.

2. Identify which of the following statements was probably made by a conservationist and which by an oil driller.

a. Sometimes our environment must be sacrificed for economic needs.
b. Previous drilling in similar areas has damaged the environment.

3. Explain briefly whether you feel the person who wrote the passage was objective or tried to sway you in a particular direction (toward the viewpoint of the drillers or that of the conservationists).

Answers are on page 256.

Lesson 3 Accuracy of Facts Based on Documentation

How can you tell whether "facts" presented are really facts?

Documentation is proof on paper. When you do your taxes, you are expected to provide documentation (your W-2 forms) of your earnings.

Similarly, writers should provide you with enough information to determine whether the facts they present are accurate. It is sometimes difficult to determine whether data are accurate; it is easier to note whether the author has included data at all. When trying to assess whether facts are supported by documentation, ask yourself: Does the writer present unsupported "facts," or are the "facts" backed up?

Assessing Documentation of Facts

Consider once again the graph about environmental quality, page 33. The caption tells you exactly who was polled and how they were selected: a sample of 1,500 *National Wildlife* readers was chosen at random. Check captions and labels when trying to determine whether the writer has documented the figures presented. Ask: Who collected the data? What was measured?

Suppose the following paragraph had been added to the passage on drilling for oil in Alaska, page 60.

> Oil drillers say their rigs would not conflict with nature, a contention roundly disputed by the Sierra Club and other environmental groups. The environmentalists cite mishaps in the Prudhoe Bay Development, including toxic seepage that killed vegetation as much as 10 miles from a drilling waste pit.

In the paragraph, which "fact" is supported by documentation?

(1) Oil rigs would not conflict with nature.

(2) Oil rigs probably would conflict with nature.

The correct answer is Choice (2). The writer tells us about a specific instance where oil drilling did cause damage to nature. The writer does not document the drillers' claim that oil rigs would not conflict with nature. Does this necessarily mean that the oil riggers are wrong and the environmentalists are right? No. The writer may have chosen not to include evidence supporting the drillers' view.

Steps in Assessing the Accuracy of Facts

1. Ask yourself whether there is any evidence given to prove the "facts."
2. If there is proof, ask yourself how strong it is.
3. If there is no proof, ask yourself why it was omitted.

Lesson 3 Exercise

1. Suppose you opened your local paper and found the following headline: "Documented Sighting of Bigfoot." Suppose that the documentation turns out to be a photograph. Why might this "documentation" be insufficient to convince a scientist that Bigfoot exists?
2. If a passage begins with the statement "A major earthquake will strike California in the next 25 years," what sort of documentation do you think should follow?

3. Reread the passage about nuclear power on page 49.

Which of the following statements in the article is supported by documentation?

(1) Nuclear electricity is an economic alternative to oil.
(2) Nuclear electricity is a safe alternative to oil.
(3) Abundant energy is a must in the competitive world market.
(4) All of the above.
(5) None of the above.

Answers are on page 256.

Lesson 4 Indicate Logical Fallacies in Arguments

Why is it useful to be able to detect logical fallacies?

Suppose a child tells you, "I can make things happen by wishing for them. I know because last night I wished for a snow day, and today school was closed because of snow." You would point out the error, or fallacy, in the child's argument: Just because one event happened after another event does not mean that one caused the other.

A good, critical reader is also aware of misleading arguments in written material. On the GED, you will be asked to identify the fallacies that appear in such materials as advertisements and biased articles.

Indicating Logical Fallacies in Advertisements

Look again at the cheese advertisement on page 48. Suppose that it turned out that only two out of sixty famous chefs polled said they used Heide brand Swiss cheese. What is the fallacy in the argument that the finest chefs use the cheese? The writer of the ad has used only a couple of selected instances to make a broad statement. The statement is not exactly false, but it is misleading.

Indicating Logical Fallacies Based on Passages

Reread the passage about cancer on page 52. Which of the following statements is NOT a misleading interpretation of the data?

(1) The marker protein signals the presence of cancer.
(2) The marker protein causes the cancer.
(3) The cancer causes the marker protein.
(4) Someone dying from heart failure will have the marker protein.
(5) The protein can be used to detect all "hidden" cases of cancer.

Only Choice (1) seems to follow logically from what we are told. Eliminate Choices (2) and (3) because you know that when two things happen simultaneously, you cannot assume that one causes the other. Choice (4) uses faulty reasoning too. Both cancer and heart failure can be fatal, but it is not necessarily true that the marker protein will detect both. Be careful not to assume that one thing is connected to another just because they have something in common. Choice (5) is flawed by overgeneralization. Although the test picked up one "hidden" case, it may have missed others.

Key Questions to Detect Logical Fallacies

1. Is a misleading cause-effect relationship being stated?
2. Is a false connection being drawn?
3. Is an overgeneralization being made?
4. Are statistics being used in a misleading way?

Lesson 4 Exercise

1. Suppose you are investigating a charge that rat cages in a particular laboratory are overcrowded. You are told that "the average cage holds only two rats." How might the statement be misleading?
2. Which sentence demonstrates a generalization that is too broad?
   (1) The *World Almanac* reports it was the coldest winter in 20 years.
   (2) Several studies show that DDT can weaken osprey egg shells.
   (3) It was the year of fewest aviation accidents on record.
   (4) Because the sample of water evaporates at x degrees, we know that water under these conditions evaporates at x degrees.
   (5) Because a sample of this type of moon rock melts at x degrees, we know the melting point of all moon rock is x degrees.
3. Studies have shown that smoking increases the risk of heart disease. TRUE or FALSE: From this statement, we can conclude that those who stop smoking will have the same lowered risk of heart disease as those who have never smoked.

Answers are on page 256.

Chapter 5 Quiz

1. Which of the following three test items is an evaluation item?
a. According to the passage, what is the effect of a stroke?
b. Which of the statements about strokes found in the passage is best supported by evidence?
c. Based on the passage, choose the person from the list that follows whose symptoms suggest he has had a stroke.

2. Suppose you want to know how safe a certain medication is. Which source is likely to be the most neutral?
(1) the pharmaceutical company that produces the drug
(2) the Food and Drug Administration
(3) the Chamber of Commerce in the town where the drug is made
(4) advertisements in medical journals
(5) the labor union leader in the pharmaceutical company

Items 3 to 7 are based on the following passage.

Are sharks and other members of their family immune to cancers? Do their bodies contain a chemical that might be useful for treating malignant tumors in humans? These are questions Dr. Carl Luer is trying to answer. He says he is making progress. He says he has documented that sharks are strangely resistant to tumors caused by one of the world's most potent carcinogens. He thinks their apparent resistance is probably related to something in their cartilage. He also thinks that with time and money the mystery antitumor chemical can be produced in quantities large enough to help humankind.

3. Complete the statement: Although Dr. Luer claims to have found evidence that ____________, the writer summarizing the research has not told us what the documentation actually is.

4. TRUE or FALSE: The passage provides supportive evidence for Dr. Luer's belief that the sharks' cartilage holds the key to their resistance to tumors.

5. Suppose you want to judge whether a scientist had uncovered enough data to justify coming up with a particular conclusion. Is it better to read (a) a concise summary of her work or (b) the scientist's own report. (Choose one)

6. According to the last line in the passage, the scientist is assuming that ____________ and ____________ will have the same response to certain chemicals.

7. Is there enough information in the passage to decide whether there is a fallacy in the scientist's reasoning that the substance in sharks could help treat tumors in humans? Explain your answer.

---

8. Suppose you are examining the following three pieces of written material:

a. a chart comparing efficiency of various car engines
b. an advertisement for a particular car
c. a letter to the editor of an automotive magazine

In which case or cases do you need to think about the values of the person who wrote the material, before accepting it as true?

Items 9 and 10 are based on the following figure.

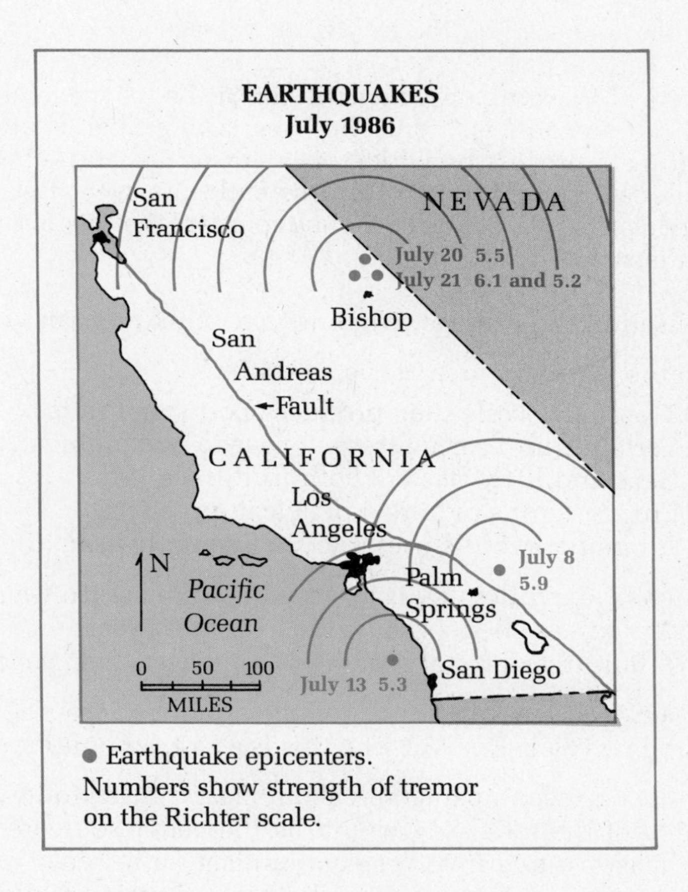

9. Does the graph provide data that adequately support the following statement?

California had five major quakes in the month shown.

10. Is there a logical fallacy in the following statement:

The quakes were caused by the fact that it was midsummer.

Answers are on page 256.

Unit I Test

Items 1 to 5 are based on the following passage.

Researchers at Washington University in St. Louis speculate that trees communicate in some fashion. They observed a phenomenon known as *mast fruiting*, a process by which oaks, hickories, and other fruit-bearing trees produce a bumper crop of acorns and nuts every few years. The trees seem to conspire in a superreproductive effort that can carpet the gound beneath them with nuts.

1. As used in the passage, what do you think *bumper crop* means?

2. What is the main idea of the selection?
(1) Trees talk while they produce acorns and nuts.
(2) Certain fruit-bearing trees appear to communicate.
(3) Oaks and hickories are both fruit trees.
(4) Bumper crops require a big clean-up effort.
(5) A bumper crop is useful to the farmer owning the trees.

3. Which of the following is an example of mast fruiting?
a. Maple trees shed extra leaves every few years.
b. Walnut trees produce extra walnuts every few years.

4. Certain fruit-bearing trees produce a bumper crop of acorns and nuts every few years. Explain why this is a fact, an opinion, or a hypothesis.

5. Suppose a second paragraph about mast fruiting follows the one above. Which of the following would be the better source of documentation that mast fruiting involves communication between trees?
a. observations by a farmer who has cultivated fruit-bearing trees for the past 50 years.
b. precise records kept by research workers of the numbers of nuts and acorns produced over a six-year period

Items 6 to 10 are based on the following figure.

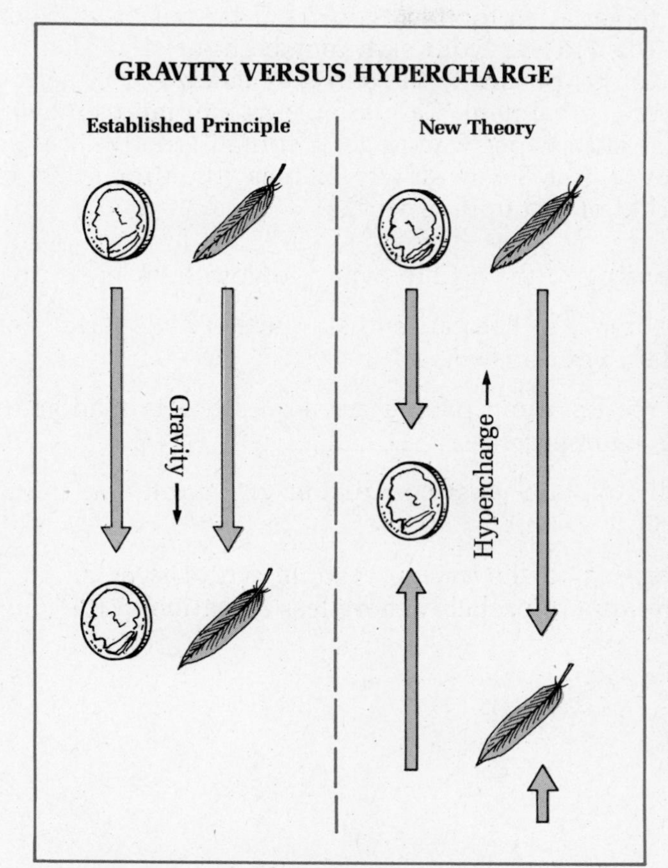

A new theory proposes that a fifth force called hypercharge pushes up against falling objects, working against the force of gravity. The force of the hypercharge is a function of the mass and the atomic composition of a given object; it is greater for a copper coin than for a feather. Thus, if a feather and a penny were dropped through a vacuum the feather would fall slightly faster than the coin. That contradicts an established principle, shown by Galileo at the Tower of Pisa, asserting that they would fall at the same rate.

6. Complete the statement: A new theory contradicts Galileo's belief that a feather ______________.

7. According to both old and new theories, what condition must be met?

(1) The theory holds true only for a copper coin and a feather.
(2) The objects must be dropped through a vacuum.
(3) The objects must be dropped from the Tower of Pisa.
(4) There must be no wind blowing.
(5) The sun must be shining.

8. According to the diagram illustrating the new theory, which would fall faster, a person weighing 250 pounds or a person weighing 110 pounds (having less mass)?

9. The new theory states that the feather would fall faster than the coin because of a force pushing up against the force of gravity.

Is the above statement a fact or a hypothesis? Why?

10. Explain what supportive evidence would be needed to prove decisively that Galileo's theory was incorrect.

Items 11 to 15 are based on the following passage.

If you try to get a tan too fast, your skin cannot produce enough melanin to protect itself. Thus, your skin may be injured and become red or blistered. Overexposure to the sun may cause a very serious burn that needs medical treatment. Careless use of sun lamps may also cause severe skin burns. The proper way to get a suntan is to limit exposure to about 20 minutes at first. Then, slowly increase the time spent in the sunlight over a period of a couple of weeks.

11. According to the article, what do you think melanin is?

12. TRUE OR FALSE: The cause of sunburn is long periods of time in the sun. Explain your answer.

13. The reason some people get a sunburn is similar to the reason that some hamburgers ____________.

14. Briefly explain what conclusion you think you should draw from the article.

15. According to the article, people with larger amounts of melanin in their skin are probably more/less sensitive to the sun. (Choose one.)

Answers are on page 257.

UNIT I TEST

Performance Analysis Chart

Directions: Circle the number of each item that you got correct on the Unit I Test. Count how many correct items there are in each row. Write the amount correct per row as the numerator in the fraction in the appropriate "Total Correct" box. (The denominators represent the total number of items in the row.) Write the grand total correct over the denominator **15** at the lower right corner of the chart. (For example, if you got 12 items correct, write *12* so that the fraction reads *12*/**15**.) Item numbers in color represent items based on graphic material.

ITEM TYPE	ITEM NUMBER	TOTAL CORRECT
Comprehension (page 31)	1, 2, 6, 7, 11	**/5**
Application (page 39)	3, 8, 13	**/3**
Analysis (page 46)	4, 9, 12	**/3**
Evaluation (page 57)	5, 10, 14, 15	**/4**
		/15

The page numbers in parentheses indicate where in Unit I you can find the specific instruction about the types of questions you encountered in the Unit I Test.

UNIT

II

Foundations In Science

In this unit, you will study the four areas of science that are covered on the Science Test. The lessons will give you practice in applying your reading skills to science materials.

The Foundations in Science unit begins with a chapter explaining what science is. Each of the next four chapters covers one of the content areas of science that appear on the GED Test: biology, earth science, chemistry, and physics. The last chapter shows how the various branches of science interact with each other to interpret phenomena.

The six chapters in Unit II are divided into lessons. Each lesson ends with an exercise made up of questions that require you to think the way you will have to when you take the GED. Each chapter ends with a quiz that covers or draws on the material presented in the chapter. The Unit II Test ends the unit. It has 33 items, the same number as in the Predictor Test in this book. About half of the items are multiple-choice.

This unit will help you develop a manner of thinking that will improve your ability to understand and interpret science materials. Your increased ability should help you perform better on the GED Science Test. Use the chart on the next page to record your progress as you work through the unit.

UNIT II PROGRESS CHART
Foundations in Science

Directions: Use the following chart to keep track of your work. When you complete a lesson and have checked your answers to the items in the exercise, circle the number of questions you answered correctly. When you complete a Chapter Quiz, record your score on the appropriate line. The numbers in color represent scores of 80% or better.

Lesson	Page	CHAPTER 1: **Introduction to Science**	
1	73	Science and the Scientific Method	1 2 3 4 5
2	74	The History of Science	1 2 3
3	76	Modern Science	1 2 3 4
4	77	Experimenting	1 2 3 4 5
5	79	Scientific Tools and Measurements	1 2 3 4 5
	81	Chapter 1 Quiz	1 2 3 4 5 6 7 8 9 10

Lesson	Page	CHAPTER 2: **Biology**	
1	83	Cells–The Basic Units of Life	1 2 3 4 5
2	86	Cells and Energy	1 2 3 4 5
3	89	The Reproduction of Cells	1 2 3 4
4	90	Genetics and Heredity	1 2 3 4 5
5	93	Characteristics and Systems of Organisms	1 2 3 4 5
6	95	The Diversity of Life	1 2 3
7	97	Living Things and Their Environment	1 2 3
8	98	The Energy Cycle	1 2 3 4
9	100	Behavior	1 2 3
10	101	Evolution	1 2 3
	102	Chapter 2 Quiz	1 2 3 4 5 6 7 8 9 10

Lesson	Page	CHAPTER 3: **Earth Science**	
1	106	Earth's Location in Space	1 2 3 4
2	108	History of the Universe and Earth	1 2 3
3	110	Structure of Earth	1 2 3
4	111	The Changing Earth	1 2 3 4
5	113	Earth's Water	1 2 3 4
6	115	The Atmosphere	1 2 3 4
7	117	Weather and Climate	1 2 3 4
8	119	Energy and the Environment	1 2 3
	120	Chapter 3 Quiz	1 2 3 4 5 6 7 8 9 10

Lesson	Page	CHAPTER 4: **Chemistry**	
1	123	Properties of Matter	1 2 3
2	124	Structure of Matter	1 2 3 4
3	126	Elements and Compounds	1 2 3 4
4	130	Mixtures and Solutions	1 2 3 4
5	132	Chemical Reactions	1 2 3 4
6	134	Acids, Bases, and Salts	1 2 3
7	136	Energy Changes in Chemical Reactions	1 2 3 4
8	137	Radioactivity	1 2 3 4
	139	Chapter 4 Quiz	1 2 3 4 5 6 7 8 9 10

Lesson	Page	CHAPTER 5: **Physics**	
1	142	Energy, Work, and Power	1 2 3 4
2	143	Motion	1 2 3
3	145	Heat and Mechanical Energy	1 2 3 4
4	146	The Nature of Waves	1 2 3 4
5	148	Behavior of Light	1 2 3 4
6	151	Electricity	1 2 3
7	152	Magnetism	1 2 3
8	154	Nuclear Energy: The Energy of the Atom	1 2 3
	155	Chapter 5 Quiz	1 2 3 4 5 6 7 8 9 10

Lesson	Page	CHAPTER 6: **Interrelationships Among The Sciences**	
1	158	Introduction to Interrelationships	1 2 3 4
2	159	Space Exploration	1 2 3
3	161	Applying Science to the Problem of Air Pollution	1 2 3 4
	163	Chapter 6 Quiz	1 2 3 4 5 6 7 8 9 10

Chapter 1

Introduction to Science

Objective

In this chapter you will read and answer questions about

- What a scientist does
- The scientific method
- The history of science
- The practice of science today
- Experimenting
- The tools and measurements of scientists

Lesson 1 Science and the Scientific Method

What is the job of a scientist?

Have you ever asked why or how something happens? Did you ever come up with a new idea about how to do something, and then put it into practice? Have you ever solved a problem by trying several solutions until you found the one that worked best? Chances are you have done at least one of the above, all of which are part of a scientist's job.

A **scientist** uses facts to better understand the world in which we live. He or she may uncover truths about nature that no one else has discovered. The scientist then puts these facts together to reach a larger understanding, called a **theory.**

The process a scientist usually follows to formulate a theory is known as the **scientific method.** The scientific method consists of six basic parts, which may be used in any order.

1. A problem is stated.
2. Information is gathered.
3. An answer to the problem, called a **hypothesis,** is suggested.
4. An **experiment** is performed to test the hypothesis.
5. **Data** from the experiment are recorded and **analyzed.**
6. Conclusions about the problem are reached.

At this point, the scientist may either form a theory or decide to search for further evidence.

In this way, scientists have learned many things, from how to put people in outer space to why some of us have blue eyes and why some of us have brown eyes. Can you think of a scientific discovery that has affected your life?

Lesson 1 Exercise

Items 1 to 5 are based on the following passage.

Alice wants to find out which absorbs a spill better, a dry towel or a wet one. She thinks that a dry towel is better because it has more dry fibers for absorption. To find out, she pours one-quarter cup water in two different places on the floor. Then she places a wet cloth on one spot and a dry one on the other. The dry cloth soaks up nearly all the water, while the wet cloth leaves one-eighth cup water behind.

1. State Alice's problem.

2. State Alice's hypothesis.

3. How does Alice carry out her experiment?

4. What conclusion may be drawn from this experiment?

5. Based on the data, how would you expect a partially dry cloth to absorb when compared to the other two?

Answers are on page 258.

Lesson 2 The History of Science

People have always been curious about the world around them, but how did science come about?

The history of science probably begins with the history of the human race. Early people desired to know and explain their **environment.** They often made up explanations that fit the facts as they knew them. For example, lightning acted like a weapon, destroying whatever it hit, so the Greeks believed it was the sword of a god, Zeus.

Like people today, people many years ago often had practical reasons for wanting to know more about their environment. The Egyptians studied the heavens and developed a calendar so that they could predict the time at which the Nile River would flood. This helped them to plan when to harvest their crops.

It is difficult to tell exactly when science began. One of the earliest sciences was probably **astronomy,** the study of the planets and stars. Early scientists were also interested in the study of plants and animals, which later evolved into **biology.** Another area of study was natural substances, such as minerals. **Chemistry** was born as the first people mixed, heated, and experimented with these materials.

As early civilizations developed, such peoples as the Egyptians, Greeks, Chinese, and Arabs did research in medicine, mathematics, astronomy, **physics,** and agriculture. They recorded observations and acted on them to figure how best to plant and cultivate crops or to weave clothing. Romans, for example, are said to have poured vinegar on the ground to test its strength. The soil contained carbonates that reacted with the vinegar to form bubbles. The Romans were using a chemical reaction to test the strength of an acid (vinegar), although they wouldn't have called it that.

The actual scientific process of doing experiments and analysis is thought to have begun several hundred years ago. At the end of the sixteenth century, the physical scientist Galileo was one of the first to conduct time experiments and to use measurement in a systematic way. A hundred years later, Newton formed theories and tested them against the alternatives in much the same way that science is conducted today. In this way, he formulated the law of universal gravitation and three laws of motion that hold true in most situations even today. He used some of his ideas to describe by a simple formula the motion of the moon around Earth. Other early scientists laid the foundation whereby scientists today make generalizations about observations and test the generalizations by experimentation.

Lesson 2 Exercise

1. Why is it hard to pinpoint exactly when science began?

2. Ptolemy believed that Earth was the center of the universe, and that the planets and the sun revolved around Earth. Of which field of science was Ptolemy an early practitioner?

3. In the mid-1600s, Robert Hooke examined cork and other tissues with a microscope. He described the boxlike structures he saw as cells. Hooke was one of the first users of the microscope and one of the first to study cells. Use this example to evaluate the effect available equipment has had upon scientific discoveries.

Answers are on page 258.

Lesson 3 Modern Science

What are the major fields of science today? Are those fields rigid categories?

The body of scientific knowledge has become so large that no one person could expect to master all of it. Scientists today usually limit themselves to a certain field of study. In addition, they often have a specialty within that field.

Below are the major fields of science.

(1) biology: the study of life

(2) earth science: the study of Earth, the other planets, the stars, and the forces that act upon them

(3) chemistry: the study of what substances are made of and how they change and combine

(4) physics: the study of different forms of energy and matter and how they behave

The major fields encompass many areas of study. A biologist may study anything from the structure of corn plants to how the eye of a fish works. A chemist may study chemical fertilizers, plastics, or the composition of a living cell. A physicist may work with motion, heat, nuclear energy, or electricity. An earth scientist may examine the movements of the planets or the cause of an earthquake.

At times, the different fields overlap. For instance, a scientist working with the chemistry of living things is called a biochemist. A physicist may discover an application of energy, such as X-rays, which can be used in the field of medicine. It is important to understand that science is not made up of rigid categories. Many important discoveries have been made by people working in overlapping areas of science.

Lesson 3 Exercise

1. Visible light consists of electromagnetic waves that move through air or space at right angles to each other. The study of visible light is part of the field of

(1) biology
(2) earth science
(3) chemistry
(4) physics

2. The study of animals is part of the field of
   (1) biology
   (2) earth science
   (3) chemistry
   (4) physics

3. Volcanoes are formed when molten rock erupts from the ground. The study of volcanoes is part of the field of
   (1) biology
   (2) earth science
   (3) chemistry
   (4) physics

4. Water is composed of two atoms of hydrogen and one atom of oxygen. The study of water's composition is part of the field of
   (1) biology
   (2) earth science
   (3) chemistry
   (4) physics

Answers are on page 258.

Lesson 4 Experimenting

What is an experiment?

Experimenting is one of the chief ways of learning scientific principles and generalizations. Nearly everything that happens in nature involves many processes and forces at work. To understand the whole, scientists try to isolate each process and to then study it.

There are many different kinds of experiments. Some involve living things, such as plants and animals. Some involve chemicals and specialized equipment. Others involve such forces as magnetism and electricity.

Most experiments fall into three major categories. An experiment may measure one **variable** under different conditions. For example, a scientist might observe what happens to a cube of ice when kept at 0° Celsius, 50° Celsius, and 100° Celsius.

An experimenter may also try to figure out how the structure of something affects processes going on in that structure. For example, after one scientist discovered that a plant's stem is made up of fluid-filled vessels, a second scientist designed an experiment to find out how the fluid moved within the stem.

A third type of experiment is one that involves the search for something never before identified. Usually the experimenter is testing a theory. He or she is looking for something that the theory predicts is there.

All experiments have one thing in common. They are designed so that all properties remain constant except for the variable(s) being studied. This means that the scientist must control the experiment, so that the variable under study is the only thing affecting the outcome of the experiment. For example, in studying plant growth, the scientist must make sure that each plant has the same amount of light, water, soil, and so on.

In the real world, it is very hard to control every variable except for the one that you are studying. This is why experiments are often conducted in a **laboratory** under controlled conditions.

Lesson 4 Exercise

Items 1 to 5 are based on the following passage.

To prove that plant leaves give off water, Carlos placed a plastic bag over a plant and sealed it tightly around the stem of the plant. This shut off the soil from contact with the air in the bag. The next morning he found droplets of water inside the bag. The next night he set up another experiment exactly like the first—a plant pot, a plastic bag, soil, etc., but without a plant. This time water appeared on the inside of the bag with the plant in it and not inside the other bag.

1. Which variable(s) stayed the same both nights?

2. Which variable(s) did Carlos change?

3. After the first night, could Carlos be sure that the water did not come from the air trapped in the bag?

4. After the second night, could Carlos be sure that the water in the bag with the plant came from its leaves?

5. Which of the following would cause the greatest error in the experiment?

(1) using different soil in the two setups
(2) not sealing the bag tightly around the plant
(3) placing the plant setup in a different part of the house from the setup without the plant
(4) using dry soil for both setups
(5) watering the plant before beginning the experiment

Answers are on page 258.

Lesson 5 Scientific Tools and Measurements

What tools and measurements do scientists use?

Like other workers, scientists use special tools to do their jobs. Scientists use tools such as the **microscope** and **telescope** to extend their senses. Scientists use instruments such as clocks, **meters,** and rulers to take measurements. During an experiment, scientists use equipment such as a test tube to isolate the effect they are studying. Finally, many scientists today use **calculators** and **computers** to help them to store and to analyze data.

Scientific tools must be accurate and reliable. An instrument for measuring must take exact measurements. Tools should be kept clean and in good repair. Although this also applies to tools used for everyday tasks, it is especially important in science. An inaccurate instrument may give incorrect results, leading the scientist to form an incorrect conclusion.

Scientists use a system of measurement known as the **metric system.** The metric system is easy to use because units of measure are divided into groups of ten. Each unit in the metric system is ten times smaller or larger than the next unit.

The **meter** is the basic unit of length in the metric system. The **liter** is the basic unit of **volume.** The **kilogram** is used to measure **mass,** and the **Celsius scale** is used to measure temperature.

COMMONLY USED METRIC UNITS

LENGTH

Length is the distance from one point to another.

A meter is slightly longer than a yard.

1 meter (m) = 100 centimeters (cm)

1 meter = 1000 millimeters (mm)

1 meter = 1,000,000 micrometers (μm)

1 meter = 1,000,000,000 nanometers (nm)

1 meter = 10,000,000,000 angstroms (Å)

1,000 meters = 1 kilometer (km)

MASS

Mass is the amount of matter in an object.

A gram has a mass equal to about one paper clip

1 kilogram (kg) = 1000 grams (g)

1 gram = 1000 milligrams (mg)

1000 kilograms = 1 metric ton (t)

VOLUME

Volume is the amount of space an object takes up.

A liter is slightly larger than a quart.

1 liter (L) = 1000 milliliters (mL) or 1000 cubic centimeters (cm)

TEMPERATURE

Temperature is the measure of hotness or coldness in degrees Celsius (°C).

0°C = freezing point of water

100°C = boiling point of water

Lesson 5 Exercise

1. Which of the following is probably not the tool of a scientist?
   (1) microscope
   (2) stopwatch
   (3) computer
   (4) measuring cup
   (5) micrometer (an instrument that measures to the nearest one-hundredth of an inch)

2. Which has the greatest length?
   (1) a nanometer
   (2) a millimeter
   (3) a centimeter
   (4) a meter
   (5) a kilometer

3. Which is larger, a liter or two quarts?

4. Which has the least mass, a feather with a mass that equals one gram or a rock with a mass that equals one-half gram?

5. Would water that measures 100°C (degrees Celsius) burn your skin?

Answers are on page 258.

Chapter 1 Quiz

Items 1 to 6 are based on the following passage.

Benjamin was interested in knowing whether different colors of cloth conduct heat differently. He chose cloths of black, blue, yellow, and white. On a sunny day, he put the cloths out on the snow. After two hours, the black had sunk 9 centimeters deep into the snow. The blue sank 6 centimeters deep, and the yellow sank 3 centimeters. The white had not sunk at all. Benjamin concluded that light cloth absorbs less light and less heat than does dark cloth.

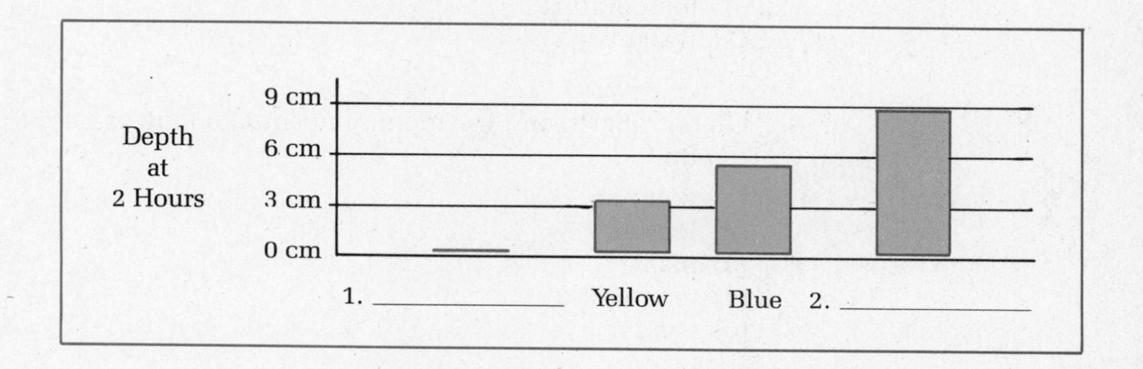

1. Fill in the blank labeled *1* under the graph.
2. Fill in the blank labeled *2* under the graph.
3. What was the variable in Benjamin's experiment?
4. Which of the three major categories (discussed in Lesson 4) does Benjamin's experiment fall into? Explain your answer.
5. What caused the black cloth to sink into the snow the farthest?
6. Based on Benjamin's conclusion, why would white or light clothing be best to wear on a hot, sunny day?

Items 7 to 10 are based on Lesson 5 and the following information.

(1) **volume:** the amount of space an object occupies
(2) **length:** the distance from one point to another
(3) **mass:** the quantity of matter in an object
(4) **temperature:** the degree of hotness or coldness
(5) **weight:** the force exerted on a mass by gravity

7. You buy a carton of milk that equals one liter. A liter is a measure of its
(1) volume
(2) length
(3) mass
(4) temperature
(5) weight

8. On both Earth and the moon you equal 60 kilograms. This is a measure of your
(1) volume
(2) length
(3) mass
(4) temperature
(5) weight

9. The outdoor thermometer reads 38 degrees Celsius. This is a measure of
(1) volume
(2) length
(3) mass
(4) temperature
(5) weight

10. A doctor tells a parent that a young boy measures about one meter from the top of his head to the bottom of his feet. This is a measure of his
(1) volume
(2) length
(3) mass
(4) temperature
(5) weight

Answers are on pages 258–259.

Chapter 2

Biology

Objective

In this chapter you will read and answer questions about

- Cells—the basic units of life
- Cells and energy
- Reproduction of cells
- Characteristics and systems of organisms
- Genetics and heredity
- The diversity of life
- Living things and their environment
- The energy cycle
- Behavior
- Evolution

Lesson 1 Cells—The Basic Units of Life

Why is a cell considered to be the smallest living thing? What are the parts of a cell?

All living things share certain characteristics that separate them from all nonliving things. Living things carry on certain activities, called **life processes,** that make them all alike in several ways. All living things use energy: plants get energy from the sun, and animals get energy from food made by plants. Living things may grow, or increase in size or mass, by using their own energy. (Nonliving things may also increase in size or mass, but only when material is added by outside forces.) Living things respond to stimuli in the environment, such as temperature, light, food, sound, and the threat of death. If necessary, living things change in order to survive. They can also reproduce themselves. (Nonliving things cannot adapt to survive and cannot reproduce.)

The basic unit of life is the cell. **Cells** are the smallest living things that exist. Although cells are usually microscopic, all the processes that characterize living organisms take place within them. They grow, they use energy, they respond to stimuli, and they reproduce themselves. Cells are also the building blocks from which all living things are made. Small organisms may be composed of only a few cells or even one cell. Large organisms are composed of many cells working together. A human being is made up of more than a trillion cells.

Cells are made of chemicals that are combined in **compounds.** Some of those compounds are familiar in everyday life, such as starches and proteins. The compounds are not alive; they work together in the structures of cells to carry out the life processes of the cells.

Each of the structures in a cell performs a specific function necessary to a life process. The cells of all organisms contain the same types of structures, although there are some differences between the cells of animals and the cells of plants (see Figure 1). Each plant and animal cell is surrounded by a cell membrane that separates it from other cells. The **cell membrane** allows essential substances into the cell and keeps unwanted substances out. Plant cells also have a tough **cell wall** surrounding the cell membrane; the cell wall gives a plant its rigidity.

Most cells contain a **nucleus,** which controls the cell's activity. The nucleus contains chemicals that are necessary for reproduction.

The cell membrane encloses the **cytoplasm,** a watery material that surrounds the nucleus. The cytoplasm contains **organelles,** the structures that carry out the cell's functions. Hundreds of organelles called **mitochondria** produce energy for the cell. Fluid-filled organelles called **vacuoles** store food particles and waste material. Other organelles are responsible for breaking food molecules down and transporting energy throughout the cell. Plant cytoplasm also contains green **chloroplasts** that capture light energy and use it to make the cell's food.

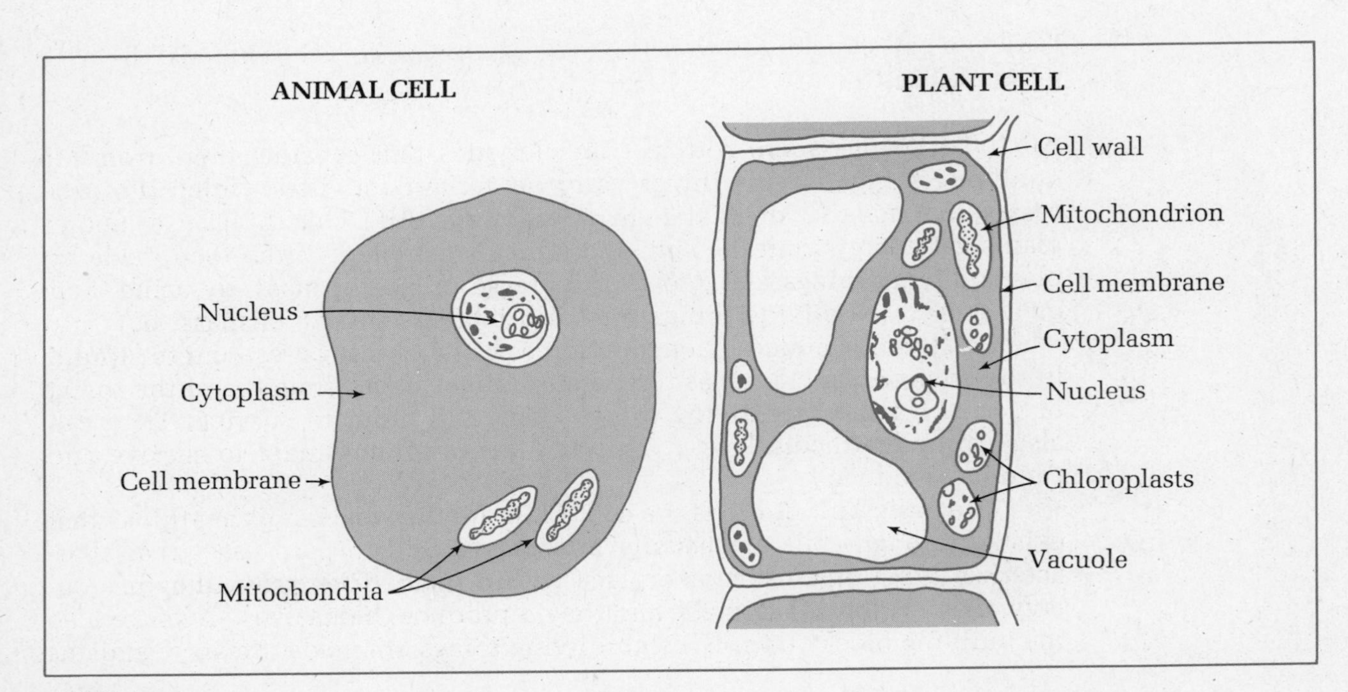

FIGURE 1

Each of the structures in an individual cell has a specialized task; together, these tasks keep the cell alive. A single-cell organism, therefore, is a self-contained entity. However, in a multicelled organism, each cell has a specialized role and works with other cells to keep the whole organism alive. The human body, for instance, contains over 100 different types of cells. Specialized cells are found in different shapes and sizes depending on their function. Muscle cells are long and thin so that they may contract and provide force. Cells that form the skin's surface are broad and flat to protect and to cover the cells beneath the surface.

Lesson 1 Exercise

1. How is the growth of a rolling snowball different from the growth of an apricot tree?

2. State whether the following is TRUE or FALSE and explain your answer: Plant cells and animal cells differ because they carry on different life processes.

3. What part of a cell performs a function similar to that of a secretary who screens calls and correspondence for an employer?

4. Why is a cell in a multicelled organism less independent than a cell that makes up a single-celled organism?

Item 5 is based on the following drawing.

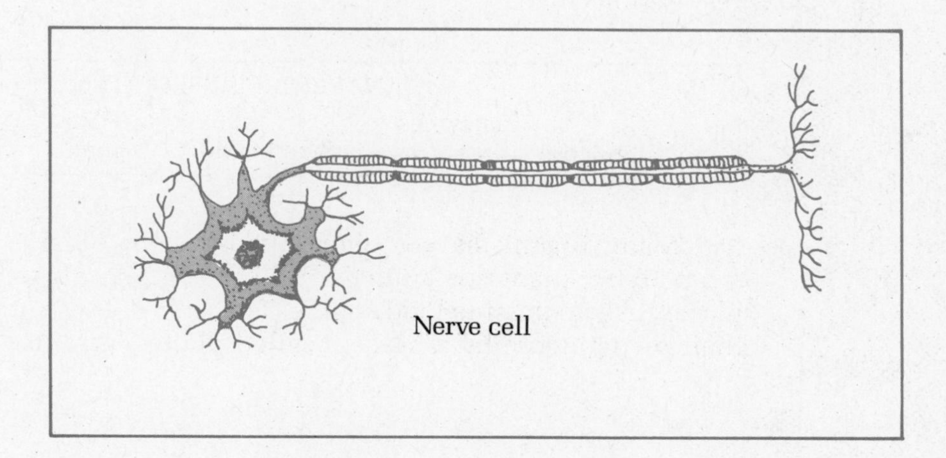

5. The function of a nerve cell is to relay messages between distant parts of the body. How is its structure related to its function?

Answers are on page 259.

Lesson 2 Cells and Energy

How do cells obtain the energy that they need to carry on life processes?

A cell needs a constant supply of energy to carry on its life processes. The most common food substance from which cells get their energy is **glucose,** a simple form of sugar. Through a series of chemical changes, cells break down glucose and thereby release energy into the cell. Chemicals called **enzymes,** which are found in a cell's mitochondria, assist in breaking down glucose for energy. The process by which energy is released from food is called **cellular respiration.**

There are two ways a cell can respire. Most organisms need oxygen for respiration to take place. Cells get the oxygen they need from air and water. Cellular respiration that uses oxygen is known as **aerobic respiration.** In aerobic respiration, most of the energy stored in glucose is released, but small amounts of water and carbon dioxide are formed by the chemical changes and must be expelled by the cell.

AEROBIC RESPIRATION

glucose + oxygen $\xrightarrow{\text{enzymes}}$ carbon dioxide + water + chemical energy

The cells of some simple organisms, such as yeast and bacteria, release energy by a process that does not use oxygen. That process is called **anaerobic respiration.** In this form of respiration, only a small amount of the energy that is stored in food is released, and more waste products are formed. The usual waste products of anaerobic respiration are carbon dioxide and alcohol. Some fermented drinks, such as beer, are a product of anaerobic respiration in yeast.

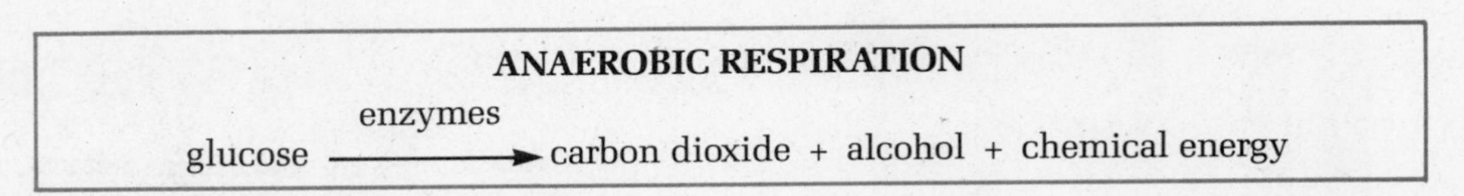

Many organisms get their glucose from outside food sources—from other living plants or animals. However, some organisms, including most plants, use no outside food source. The cells of these organisms manufacture their own glucose by a process called photosynthesis.

In **photosynthesis,** a plant absorbs energy from sunlight through a green substance called **chlorophyll,** which is contained in the chloroplasts. (It is the green of chlorophyll that gives plants their color.) The chloroplasts in the cells of a plant convert carbon dioxide, water, and the energy from sunlight into glucose. Some of the glucose produced by photosynthesis is used immediately; the rest is stored for use at night, when photosynthesis is not possible. Along with glucose, oxygen and water are produced during photosynthesis. Note that although plants then use some of the oxygen in aerobic respiration, most of the oxygen formed during photosynthesis is released into the air.

PHOTOSYNTHESIS

$$\text{carbon dioxide} + \text{water} + \text{sunlight energy} \xrightarrow{\text{chlorophyll}} \text{oxygen} + \text{water} + \text{glucose}$$

Lesson 2 Exercise

1. What substances must a cell obtain from outside itself in order for respiration to take place?
2. Many more ocean plants live near the surface of the water than live several feet below. Why do you think this is so?
3. How do plants affect air composition in a room?

Items 4 and 5 refer to the following figure and paragraph.

Tube worms live 2,600 meters beneath the sea. After discovering tube worms living near the deep ocean vents, scientists first hypothesized that the warm vent water was supporting life. Now they believe the tube worm receives nutrients from the bacteria and in turn supplies them with the raw materials they need to carry out life processes.

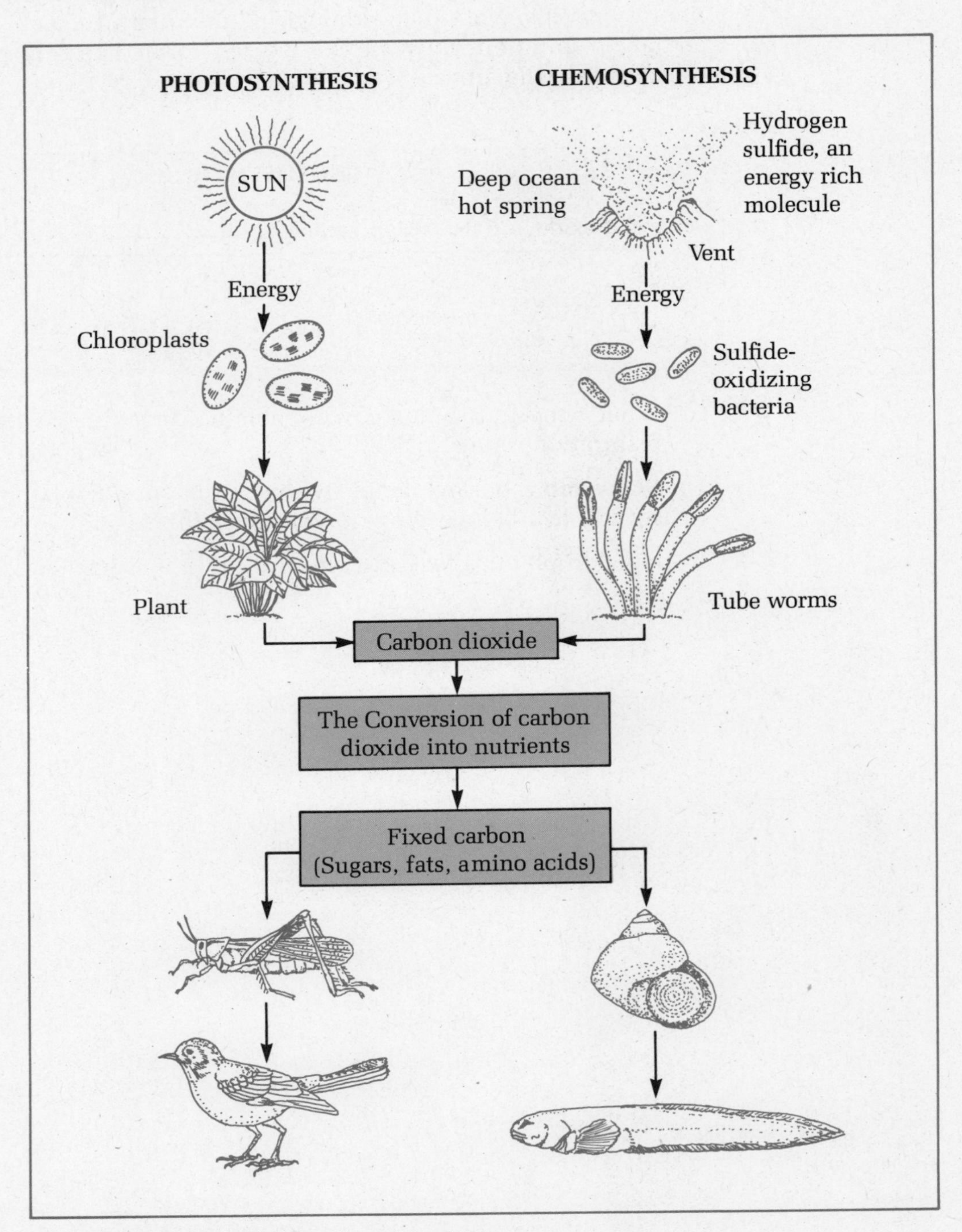

Reprinted from *Scientific American*, May 1984.

4. How is chemosynthesis both like and unlike photosynthesis?

5. Which of the following conclusions may be drawn from the figure?
   (1) Sulfide-oxidizing bacteria are replacing green plants as energy sources for most animals.
   (2) Tube worms are not capable of aerobic respiration.
   (3) Tube worms are to chemosynthesis as photosynthesis is to green plants.
   (4) Snails and fish feed on the sulfide-oxidizing bacteria.
   (5) Both chemosynthesis and photosynthesis are capable of supporting a chain of life.

Answers are on page 259.

Lesson 3 The Reproduction of Cells

How do living things grow and reproduce?

Although the cells in your body are constantly dying, you don't notice it. This is because the body's cells are constantly making new cells to take the place of dying cells. Cells within an organism, as well as single-celled organisms, reproduce by a process known as mitosis. In **mitosis,** a cell divides to form two identical cells. The two new cells have exactly the same characteristics as the original cell. In other words, a single muscle cell can divide itself into two new muscle cells. Mitosis is known as **asexual reproduction** because new cells are formed from a single parent cell.

How is the cell able to reproduce itself exactly? Within the nucleus of a cell are rodlike structures called **chromosomes.** Each chromosome is made up of a complex chemical called **DNA.** The chromosomes contain all the information that determines the characteristics of the cell. Without chromosomes, the cell would not be able to reproduce itself.

During mitosis, the chromosomes within the cell's nucleus create exact duplicates of themselves. When the cell splits apart, each new cell receives its own set of chromosomes, which form a new nucleus in each cell. Because the chromosomes in each new cell are the same, the new cells will be exactly alike. Almost all cells in an organism reproduce in this way. The only exception is the nerve cell. Once a nerve cell matures, it cannot reproduce.

The way in which cells reproduce themselves within an organism is different from the way in which an entire multicelled organism reproduces itself. Most multicelled organisms reproduce themselves through **sexual reproduction,** in which a new cell is formed from joining two parent cells. The main process in sexual reproduction is **meiosis.**

Meiosis is a division of the cell similar to mitosis, except the chromosomes do not duplicate before the cell divides. When the parent cell splits, therefore, the two new cells are incomplete. They are known as **gametes** because they have only half the normal number of chromosomes. Sexual reproduction occurs when a gamete from a male parent unites with a gamete from a female parent, and a new cell is formed that has a complete number of chromosomes. This new cell is the fertilized egg, from which the offspring develops.

Lesson 3 Exercise

1. Which of the following organisms is most likely to reproduce asexually?
   (1) a jellyfish
   (2) a bacterium
   (3) an oak tree
   (4) a human being
   (5) a frog

2. A chicken's egg is a female gamete that is formed during meiosis. Based on this information, why doesn't a chicken's egg always hatch?

Items 3 and 4 refer to the following figure.

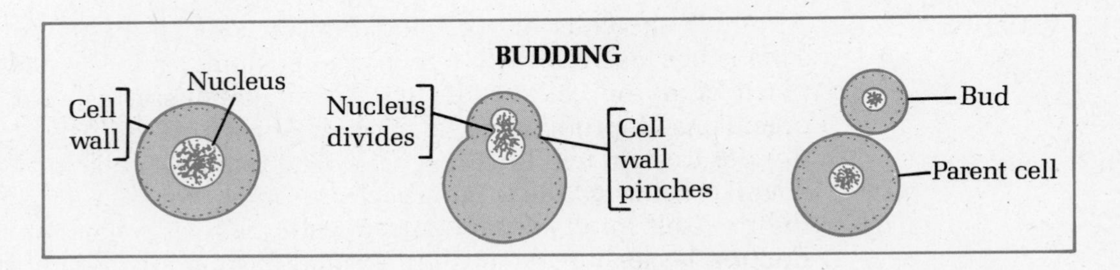

3. Budding is which type of reproduction—asexual or sexual?
4. What is the main difference between a bud and a gamete?

Answers are on page 259.

Lesson 4 Genetics and Heredity

How are characteristics passed on to offspring?

Most single-celled organisms reproduce through mitosis; therefore, the new organisms have the same characteristics as the parent organisms. Complex organisms that reproduce sexually, however, like plants and animals, have offspring that have different characteristics than their parents. This is because the offspring receive chromosomes from *two* parents. Thousands of characteristics, known as **genes,** are contained in the chromosomes of each parent. These genes determine all the traits of an organism, including height, sex, and hair color. When the chromosomes from two parents are combined in sexual reproduction, the offspring may have traits that are neither exactly like one parent nor exactly like the other. The study of how offspring receive traits from their parents is known as **genetics.**

One of the first genetic experiments was conducted by Gregor Mendel. Mendel noticed that certain pea plants produced only tall plants while others produced only short plants. Mendel then crossed a tall plant with a short plant. The offspring of these plants were all tall plants. Then Mendel crossed this second generation of tall plants; this time both tall and short plants were produced—75% were tall and 25% were short. How could this be explained?

Mendel offered several hypotheses that have now been confirmed by scientists who study genetics. Mendel suggested that every trait in an organism is composed of two factors—one from the set of genes given by one parent and the other from the set of genes given by the other parent. For example, a pure tall plant contains two genes for tallness. A pure short plant contains two genes for shortness. When the two plants are crossed, the offspring receive one gene for shortness and one gene for tallness. Remember that in meiosis, parent cells split into gametes that have only half the normal set of chromosomes, and therefore have only one gene to contribute for each trait. It would seem then that the offspring could be either short *or* tall. However, Mendel's experiments showed that those offspring with one short and one tall gene were *always* tall. Mendel concluded that tallness is a stronger trait than shortness.

Scientists call the gene for tallness a **dominant** gene in pea plants. Whenever this gene is present in the chromosomes of the offspring, it dominates or controls the development, and the plant develops as a tall plant. The gene for shortness is labeled a **recessive** gene. If it combines with a dominant gene, the recessive trait will not show up, although the gene will remain in the plant's chromosomes.

	T	T
s	Ts	Ts
s	Ts	Ts

FIGURE 2

Figure 2 is a chart representing the first generation of Mendel's pea plants. Along the top of the chart are shown the genes of the pure tall plant. It has two genes for tallness (represented by a "T"). Along the side are shown the genes for the pure short plant. It has two genes for shortness (represented by an "s"). The possible combinations of these genes in the offspring (after meiosis and sexual reproduction) are shown in the boxes. Note that every possible combination has a tall gene. Because the tall gene is a dominant one, any offspring from these parent plants will be a tall plant.

Figure 3 shows the second generation of Mendel's pea plants. Both of the parent plants now contain a tall gene and a short gene. The possible combinations of these genes in the offspring are now quite different. As you can see, three of the four possibilities will have the dominant trait of tallness. However, there is a 1:4 chance that two recessive genes would combine. This would cause the resultant offspring to be *short* even though both of the parent plants were tall.

	T	s
T	TT	Ts
s	Ts	ss

FIGURE 3

The way such characteristics are passed on from one generation to the next is called **heredity,** because the traits are inherited from the offspring's parents. In humans, heredity is far more complex than in pea plants, but the basic principles are the same. Heredity explains why certain traits are characteristic of one family, and why children may look more like one parent than like the other.

Lesson 4 Exercise

1. Explain why a single individual cannot have more than two genes for a particular trait.
2. In humans, brown eyes are dominant and blue eyes are recessive. Make a square chart showing the possible eye colors of children of a mother who has two genes for brown eyes and a father who has two genes for blue eyes.
3. A man who has one gene for blue eyes and one for brown eyes marries a woman with blue eyes. What are the chances their children will have blue eyes?

Item 4 is based on the following chart.

BLOOD TYPE GENES	ACTUAL BLOOD TYPE
AA or AO	A
BB or BO	B
AB	AB
OO	O

4. Three genes for blood type exist in human beings—A, B, or O. The chart shows how they express themselves in various pairings. Which of the following conclusions about the blood types may be drawn from the chart?

   (1) A, B, and O are all dominant traits.
   (2) Only A and B are dominant traits; O is recessive.
   (3) Only A is a dominant trait; B and O are recessive.
   (4) Only A and O are dominant traits; B is recessive.
   (5) Only B is a dominant trait; A and O are recessive.

5. How does our knowledge of genetics assist us in breeding plants for our use?

Answers are on page 259.

Lesson 5 Characteristics and Systems of Organisms

How are complex organisms structured?

We have seen how cells carry out the life processes, and how they are structured in order to perform the tasks that are necessary to these processes. Complex, multicelled organisms, such as plants and animals, require exactly the same processes in order to sustain life. Like the cell, they must breathe, obtain nutrition, excrete waste products, and reproduce. But because they are responsible for sustaining not just one cell, but the many billions of cells that make up the entire organism, the ways in which they carry out these processes are very different and far more complicated.

A plant has specialized sections to help it perform all the functions necessary to life. It absorbs water and minerals from the soil through its **root system.** It holds itself upright, toward the sun, and connects the root system to its upper portion with its stem. The **leaves** of a plant absorb carbon dioxide from the air and sunlight to convert to energy in photosynthesis. Water and oxygen are excreted into the air through the leaves. Tiny tubes known as **capillaries** transport water and food from one part of the plant to another. It reproduces itself by producing male and female gametes—pollen and egg cells—within its **flowers.**

These sections are made of millions of specialized cells, each of which must be supplied with food, water, and oxygen. Working together, these parts of the plant furnish each cell with what it needs, and at the same time help to keep the entire organism alive.

Animals, like plants, require the basic necessities of life—food, air, and water. But because their bodies are even more complex than those of plants, their cells and parts are even more specialized.

The cells themselves are organized into **tissues,** according to the jobs they do. For instance, all the cells that perform such functions as lining and protecting different parts of the body are called **epithelial tissue.** Other types of tissue are **muscle tissue,** which gives the body its ability to move, **nerve**

tissue, which makes the body aware of what is around it, and **connective tissue,** which connects muscles to bones and makes up the fabric of the bone itself. All cells within a tissue are similar to each other, but differ in form and function from cells in other types of tissue.

Different groups of tissue may work together to do a certain job. They are organized into structures called **organs.** The heart, for instance, has muscle tissue to make it pump, nerve tissue to connect it to the brain, epithelial tissue to protect it from bruising and infection, and connective tissue to hold it in place.

Working together, organs are grouped into systems, which are the highest levels of organization within an organism. A **system** is responsible for providing the whole body with a vital life function. For instance, the function of the **digestive system** is to break down food into energy usable by the cells of the body. The mouth, stomach, and intestines are the major organs in the digestive system. None of the organs in the system can perform the function of digestion by themselves; each organ is simply a part of the process. The mouth breaks down food into smaller particles and secretes saliva to begin the digestive process. In the stomach, enzymes and acid are added to the food mass to make it liquid. The food is then passed into the small intestine, which is lined with cells that produce digestive juices. The small intestine then absorbs the digested food and passes it into the bloodstream. Solid materials that can't be digested or used by the body pass into the large intestine, where they are stored until elimination takes place.

The digestive system is just one of many complex systems found in animals. Other life processes are carried out by separate systems. Oxygen is taken from the air and carbon dioxide released into the air by the **respiratory system.** Food, oxygen, and other chemicals are carried by the blood to every cell in the body. The vessels carrying the blood and the heart that pumps the blood make up the **circulatory system.** The circulatory system also transports waste materials from the cells to other parts of the body where they are eliminated. The process of elimination of solid and liquid waste is the function of the **excretory system.**

Animals also have **muscular systems,** which enable the body to move, and **skeletal systems,** which support the body and protect the body's internal organs. The complexity of an animal's body systems requires that another system coordinate all the different activities that take place. The **nervous system,** whose main organ is the brain, provides such coordination.

Lesson 5 Exercise

1. Arrange the following words in order from the highest to the lowest form of organization: cell, organ, system, tissue.
2. What is the purpose of a system?
3. What system in the human body performs the same function as capillaries in a plant?

4. Why do you think a plant lacks a digestive system?

5. Polio is caused by a virus that attacks the motor neurons in the spinal cord and/or the brain stem. The motor neurons control various kinds of movement. Which of the following systems is least likely to be affected by polio?

(1) circulatory
(2) muscular
(3) nervous
(4) respiratory
(5) skeletal

Answers are on page 260.

Lesson 6 The Diversity of Life

How do scientists distinguish between life forms?

From microscopic, single-celled organisms to human beings, there are literally millions of different life forms on Earth. All of them share certain characteristics which have been discussed in earlier lessons. In order to better study and to understand these forms, however, scientists must examine the differences among them. To help them in this task, scientists try to identify and name each different organism. This process is known as **classification.** See the following classification chart.

THE FIVE KINGDOMS

KINGDOM	DESCRIPTION	EXAMPLE
1. Monera	One-celled Have no nuclei	Bacteria, blue-green algae
2. Protista	One-celled Have nuclei	Protozoans, algae
3. Fungi	Multi-celled Have cell walls Cannot make food	Yeast, molds, mushrooms
4. Plantae (Plants)	Multi-celled Have cell walls Use sunlight and chlorophyll to make food	Seeded plants, evergreens, ferns
5. Animalia (Animals)	Multi-celled Cannot make their own food Eat other organisms	Insects, reptiles, birds, mammals

The classification system tries to reflect similarities at the same time as it reflects differences. Therefore, it provides many levels on which an organism can be classified. The broadest level of classification is the **kingdom.** A kingdom contains the greatest number of living things. Each kingdom contains living things that are alike in the most basic ways. For example, though plants vary greatly, they all use photosynthesis to make food. And although animals vary even more, they are all multicelled and cannot make their own food. For this reason, plants and animals are each grouped into kingdoms. Scientists have discovered other such basic distinctions between life forms, chiefly in the vast number of single-celled organisms. Although they disagree on the number of kingdoms into which these organisms should be classified, they all agree that these single-celled organisms are neither plants nor animals. The chart on the preceding page lists the five kingdoms that most scientists use to classify organisms.

Scientists divide kingdoms into smaller parts by finding more specific differences between organisms. For instance, the animal kingdom is divided into animals that have spinal cords and animals that do not. Animals that have spinal cords are then further divided into warmblooded animals and coldblooded animals. This system of division continues until the smallest unit of classification is reached—the species. **Species** are defined by the fact that only members of the same species can breed with each other. Scientists make no further divisions after the species.

Lesson 6 Exercise

1. In which kingdom do human beings belong? Explain your answers.

2. In what ways are some single-celled organisms neither plants nor animals?

3. Why would a dog and a cat not be considered members of the same species?

Answers are on page 260.

Lesson 7 Living Things and Their Environment

How do living things depend on each other?

You have seen how a cell survives, and how cells within a multicelled organism work together to keep the organism alive. This does not mean, however, that all organisms are self-sufficient. Most organisms depend on other organisms for survival. For instance, animals depend on plants for food and oxygen. Plants, as well as most microorganisms, depend on animals for waste material that they can convert into nutrients. Organisms depend on each other and interact with the living and nonliving things that surround them in areas of space known as **ecosystems.** An example of an ecosystem would be a forest where trees, plants, birds, insects, animals, bacteria, fungi, rocks, and soil are the components of the ecosystem. The study of the interdependence of organisms within an ecosystem is known as **ecology.**

The organisms within an ecosystem are dependent upon one another because they each provide something that the others need. In the forest ecosystem, for instance, the trees and other plants supply food and shelter for such creatures as birds, insects, and small animals like squirrels and mice. These creatures, in turn, provide food for bigger creatures such as foxes or owls. The bacteria and fungi on the ground live off dead matter, breaking it down into new soil to the benefit of the plants.

No one organism in an ecosystem dominates the others. When all organisms in an ecosystem can survive normally, an **ecological balance** exists. Organisms survive best when there is an ecological balance. When something occurs to throw off the balance, the survival of organisms is threatened.

Organisms of the same species that live in the same area are known as a **population.** If the size of a population in an ecosystem changes, an imbalance will occur. The size of a population is usually limited by the amount of food available. If a source of food begins to dwindle, all the populations that feed on that source will also dwindle until a new balance is achieved. Similarly, if a population grows too quickly, it will become too dense, or crowded. Some members of the population will be competing for the same food; those who can't get enough food will die of starvation. Again, this will occur until a new balance is reached.

An ecosystem's balance is very delicate. Even if only one population is out of balance, the entire ecosystem can break down. Floods and other natural disasters can greatly harm an ecosystem. Human interference, too, can often cause an imbalance in the environment. For instance, pesticides used in agriculture may destroy insects that are harmful to crops. But those same insects provide food for other insects and for birds, which in turn provide food for larger animals. Every animal population in the ecosystem is therefore harmed by the insecticide. Likewise, when human beings cut down forests, they destroy not only trees, but also most of the populations that depend on the trees for food and shelter. These kinds of larger imbalances are much more dangerous than minor imbalances in population size, because when they occur, there is often no way for the natural environment to regain its former balance.

Lesson 7 Exercise

1. Give two examples of interdependence in an ocean ecosystem.
2. Which do you think would create more damage to a river ecosystem—a dam built across the river or the overpopulation of a certain species of fish? Explain your answer.
3. Hippos in the Nile River of Uganda play an important role in the area's ecosystem. At night, hippos eat grass on land and then return to the water to sleep by day. The hippos' dung is fertilizer for algae and water lettuce in the Nile. Fish, insects, and snails eat the plants and small particles of dung. Larger fish, in turn, eat the smaller fish and animals, as well as the plants. Humans and other predators in the area eat the fish and the hippos themselves.

   Which of the following situations would probably cause the largest imbalance in the Nile River ecosystem?

   (1) preventing people from hunting hippos
   (2) plowing up the grass around the river
   (3) removing most of the larger fish
   (4) poisoning the snails so that most died
   (5) harvesting the water lettuce

Answers are on page 260.

Lesson 8 The Energy Cycle

How does energy flow within an ecosystem?

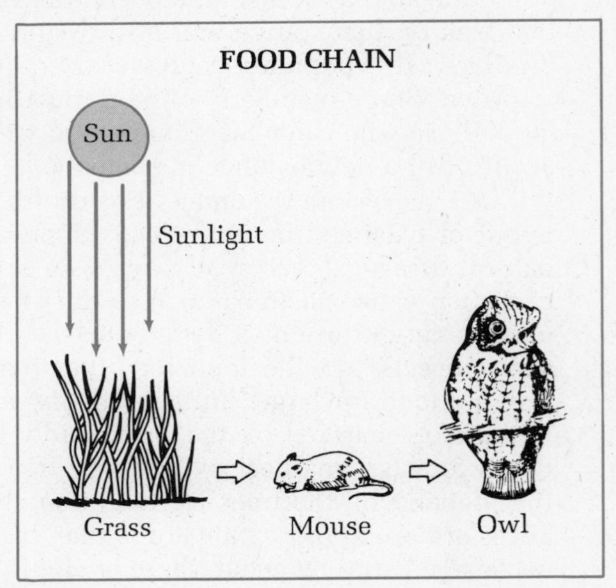

FIGURE 4

The interdependence of populations within an ecosystem is based on the flow of energy among the organisms. Every living thing must be able to obtain energy from its environment. Energy is needed to carry out all basic life processes. Where does the energy come from?

During photosynthesis, plants convert light energy into the chemical energy found in food molecules. Plants are called **producers** of food energy, because they create food where it did not exist before. Animals get energy by eating plants, or by eating those organisms that eat plants. Animals are therefore **consumers** of the food energy produced by plants. The flow of energy from the sun to producers to consumers is called a **food chain.** A simple food chain is shown in Figure 4 on the preceding page.

Food chains in an ecosystem are usually interrelated. For example, an owl may be eaten by a larger animal. Mice may be eaten by different animals, like foxes, that are found in a different food chain. Interconnected food chains make up a **food web.**

Energy is always lost at each step of the food chain. This is because the more complex organisms require more energy to keep themselves alive, and because they are less efficient at converting food into energy. This means that at each level of a food chain, less energy is available. Thus, the population size of each species is limited by the food energy available. For instance, the same ecosystem that supports a thousand shrubs and grasses may support only a hundred antelope, which in turn may supply only ten lions.

Lesson 8 Exercise

1. In a pond, insects feed on plants. Frogs feed on insects and plants, while fish feed on tadpoles, insects, and plants. Is this a description of a food chain or a food web?

2. If the example using plants, antelope, and lions in the lesson is typical of the food energy available at each level, how much energy is lost from one level to the next higher level?

3. Which of the following depicts a food chain that would provide a human being with a steak dinner?
   (1) sunlight → grasses → mice → steer → human being
   (2) sunlight → grasses → steer → human being
   (3) grasses → sunlight → steer → human being
   (4) grasses → steer → human being
   (5) human being → steer → grasses → sunlight

4. Based on the food chain shown in the text, would you expect to find more owls or mice in an ecosystem? Why?

Answers are on page 260.

Lesson 9 Behavior

What is meant by behavior? What kinds of behavior are manifested by organisms of different levels of complexity?

The way that living things respond to their environment is known as **behavior.** Behavior may be classified into different types according to its complexity. The more complex an organism is, the more complex type of behavior it exhibits.

Single-celled organisms respond to stimuli. For example, though it has no nervous system, a one-celled paramecium will still back away if poked. A plant will show a similar response: it can move toward or away from a stimulus, such as light, water, or gravity. It is such behavior that enables its roots to grow downward while its leaves stretch toward the sun.

Organisms that have nervous systems are able to exhibit behavior of increasing complexity. The three main types of behavior in such organisms are called **reflexes, innate responses** (instincts), and **learning.** A reflex is a direct response to an external stimulus. For example, cold air causes shivering, and bright light causes the pupils of your eyes to contract. Animals that have simple nervous systems are capable only of reflex action.

Innate responses are complex behavior patterns that are inherited rather than learned. They fulfill basic biological needs and are exhibited by all members of a species. Examples are web-weaving by spiders, nest-building and migration by birds, and the food-finding and courtship activities of many animals.

Learning behavior requires a memory. Not all animals are capable of all levels of learning. Learning is the ability of an organism to acquire new or changed patterns of behavior by interacting with its environment. The simplest form of learning is **habit,** a thing done often and, hence, easily. For instance, though all animals respond to stimuli, an animal can learn to ignore stimuli that are harmless. If you continually clap your hands in front of the family dog, and nothing unpleasant happens, it will learn to ignore the clapping. This is **habituation behavior. Conditioned behavior** is more complex. In conditioning, an animal learns to respond to *new* stimuli. A good example is the way guard dogs are conditioned to attack intruders. More complex still is learning by **trial and error,** in which certain forms of behavior are rewarded. A cat that knows where you store its food has probably learned this by trial and error. The most complex form of learning, one which is only exhibited by the highest forms of animals, is **insight learning.** Insight involves putting familiar things together in new ways. For instance, most animals are unable to solve puzzles. Humans, for whom insight is the basic form of learning, can do this easily.

Lesson 9 Exercise

1. Give two examples of habit behavior.

2. State whether each of the following is an example of a reflex, an innate response, or learning.
   (a) A person blinks his eyes when hands are clapped in front of his face.
   (b) Two cats mate and produce a litter of kittens.
   (c) A woman solves a math problem according to a formula.
   (d) An earth worm backs away when prodded with a needle.
   (e) A rat, after many tries, finds its way through a maze.
   (f) Bears hibernate through the winter.

3. Why is the ability to understand language an example of insight learning?

Answers are on page 260.

Lesson 10 Evolution

How and why do life forms change over a long period of time?

In order to survive, a species must be able to compete successfully for available food, water, and space. When the environment changes drastically and ecosystems are disrupted (for example, when water and food sources dry up, or when long periods of cold settle in), a species must also be able to adapt to the new conditions. Those species that cannot adapt successfully die out, or become **extinct.** Those species that survive are the ones that can adapt physically, or the ones that can change their behavior, in order to best take advantage of their new environment. Such changes are part of the process of evolution.

How does a species change over time? The process of **evolution** involves two elements, known as genetic variation and natural selection. Mutation is one type of genetic variation. A **mutation** is a sudden variation in some inheritable characteristic of a plant or animal. The genes of mutants differ from those of the parent. Sexual reproduction spreads these genes throughout the population, a process of genetic variation called **recombination.**

Natural selection, according to Charles Darwin, who developed the theory of evolution, results in the survival of the fittest. It ensures that only those genetic variations that are favorable to an organism's ability to survive and reproduce are passed on. For instance, a mouse that runs slowly is more likely to be killed by a hunting fox than a mouse that runs fast. The slow mouse will probably not live long enough to pass on its genes, but the fast mouse will survive to reproduce. Its offspring may inherit the genes that make it a fast runner. Such favorable traits accumulate in the species while the unfavorable traits gradually disappear. After many generations, the changes may result in the evolution of a new species.

Evolution is usually a very slow process because major changes in the environment tend to be drawn out over thousands of years. It may have taken an animal or a plant millions of years to evolve into its current state. However, recent harmful changes in the environment, mostly caused by the activities of humans, have occurred within a very short period of time. The organisms that are affected by these changes do not have time to adapt, and therefore, often do not survive. It is for this reason that many species have become extinct in recent years.

Lesson 10 Exercise

1. Certain species of whales are becoming extinct because of overfishing and polluted oceans. Which of the following statements best explains their failure to evolve in response to their changing environment?
   (1) The whales do not have enough time to meet those changes.
   (2) The whales lack favorable variations which would enable them to survive.
   (3) Whale genes do not mutate.
   (4) Pollution is an environmental change which no species can adapt to.
   (5) Whales produce too few offspring at a time.
2. Why do you think that insecticides such as DDT become ineffective after many years of use?
3. How do polio and smallpox vaccinations affect human evolution?

Answers are on pages 260–261.

Chapter 2 Quiz

Items 1 and 2 refer to the following passage.

Phosphorus, an important element in all biological systems, can cause problems when too much of it enters the environment. When human waste products such as detergents are deposited in lakes and streams, the blue-green algae population increases. These organisms are fertilized by the extra phosphorus. After a while, the algae form a thick layer on the top of the water. The algae near the bottom of the layer die from a lack of light. They are then decomposed by bacteria, a process that uses up a lot of oxygen. As the water's oxygen level drops, fish begin to die.

1. Complete the diagram to show how increased phosphorus in lakes and streams causes fish to decrease.

Phosphorus →	**Algae →**	**Bacteria →**	______ **→**	**Fish**
Increase	______	Increase	Decrease	Decrease

2. Which of the following statements is evidence that blue-green algae are photosynthesizing organisms?
   (1) Blue-green algae live in water.
   (2) Blue-green algae die from lack of light.
   (3) Blue-green algae are decomposed by bacteria.
   (4) Blue-green algae use phosphorus to carry out life processes.
   (5) The decomposition of blue-green algae uses up oxygen.

---

3. Normal body temperature is 98.6°F. The human body responds to infectious disease by raising the body temperature. This causes the heart to beat faster and the blood to flow faster, which helps the body fight the disease. In children, fevers lower than 102° generally are not considered dangerous. High temperatures of 104° to 106° can cause convulsions.

   If a child has a temperature that is lower than 101° and you give fever-reducing medicine, are you helping the child to recover?

Items 4 and 5 are based on the following information.

A cell moves molecules from an area of high concentration to an area of low concentration by diffusion. A cell passes waste materials out of itself by excretion. A cell moves water across its membrane by osmosis. A cell releases energy in foods through respiration. A cell takes digested food and makes new substances by synthesis.

4. When a cucumber is placed in a strong salt solution to be pickled, salt passes from the solution into the less salty cucumber. Which of the following processes is taking place?
   (1) diffusion
   (2) excretion
   (3) osmosis
   (4) respiration
   (5) synthesis

5. The liver cells put some amino acids, chemical compounds made up of carbon, oxygen, hydrogen, and nitrogen, together into chains known as protein. Which of the following processes is taking place?
   (1) diffusion
   (2) excretion
   (3) osmosis
   (4) respiration
   (5) synthesis

Item 6 refers to the following figure.

6. The leaves in the figure are each typical of their category. Which of the following statements gives the major difference between a simple leaf and a compound leaf?
   (1) A simple leaf has more veins than a compound leaf.
   (2) A simple leaf is more efficient at photosynthesis than a compound leaf.
   (3) A simple leaf consists of one leaf only; a compound leaf consists of several leaflets.
   (4) A simple leaf is broader than a compound leaf.
   (5) A simple leaf is smaller than a compound leaf.

---

7. The central office that controls all the activities of a large corporation is similar to what system of the human body? Why?

8. A fly that has brown spots and a fly that has red spots mate. Brown spots are the dominant gene. Which of the following statements CANNOT be true?
   (1) The brown-spotted fly has one red gene.
   (2) The red-spotted fly has two red genes.
   (3) There is a 50% chance that the offspring of the two flies will have red spots.
   (4) There is a 100% chance that the offspring of the two flies will have red spots.
   (5) There is a 100% chance that the offspring of the two flies will have brown spots.

Items 9 and 10 refer to the following figure.

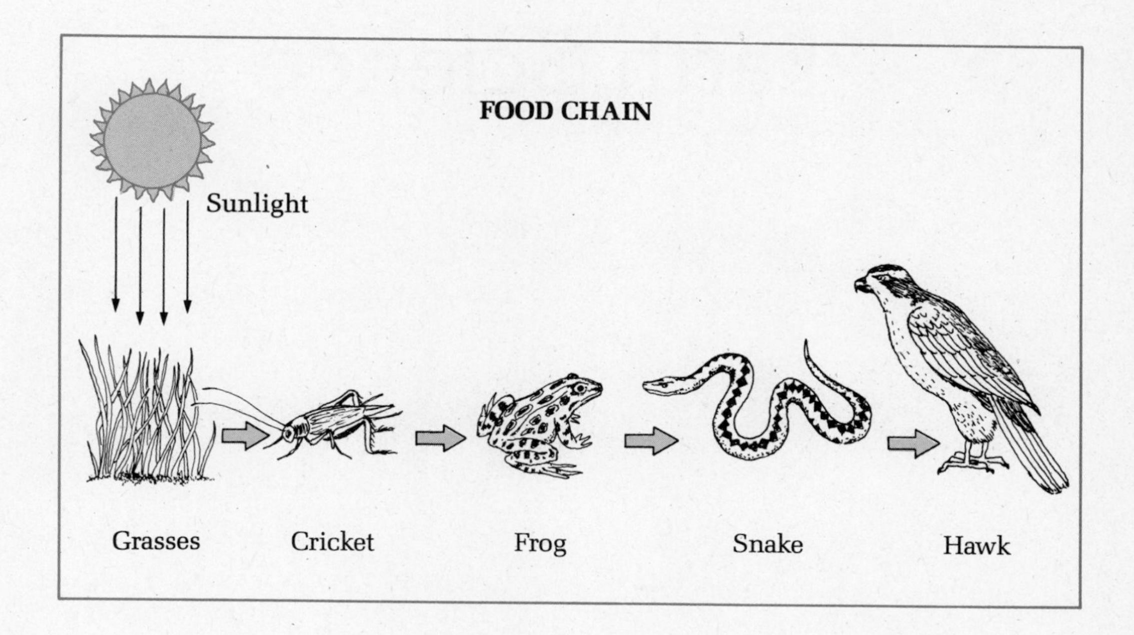

9. What would happen in the ecosystem represented by this food chain if the snake population were to be drastically reduced?

10. According to the figure, is it correct to say that plants are the ultimate source of energy for all living things? Explain your answer.

Answers are on page 261.

Chapter 3

Earth Science

Objective

In this chapter you will read and answer questions about

- the formation of the universe and our solar system
- Earth's structure and composition
- the oceans and the atmosphere
- energy from the environment
- erosion and movement of Earth's plates

Lesson 1 Earth's Location in Space

What is Earth's location in space?

Scientists' calculations of the volume of the universe show it to be much larger than anything we can imagine. It is so large that it must be measured in **light years.** One light year is the distance that light travels in one year—about 5.8 trillion miles. Most of our universe is empty space, containing no air or matter. Still, the universe contains millions of star systems—known as **galaxies**—that are themselves incredibly large. Our own galaxy, the **Milky Way,** is 100,000 light years in diameter.

The stars in a galaxy are held together by **gravity,** which is the attractive force that any object with mass exerts on the objects around it. All the stars you can see with the unaided eye are in the Milky Way. Our own sun, a star near the edge of the Milky Way, is connected to the galaxy by gravity. It is only one of an estimated 100 billion to 200 billion stars in the Milky Way.

If an object is smaller in mass than a star, and is close enough to it, the object will be caught within the star's gravitational force, and a **solar system** will be formed. Our solar system consists of the sun, nine planets and their moons, and a band of small chunks of matter called the **asteroid belt** (see Figure 1).

Each planet revolves around the sun in a path known as an **orbit.** If no force acted upon them, the planets would move through space in a straight line. But the steady force of the sun's gravity changes their paths to a closed curve. The orbit of a planet around the sun is an **ellipse,** which is similar in shape to an oval. Earth takes one year to make a complete orbit around the sun.

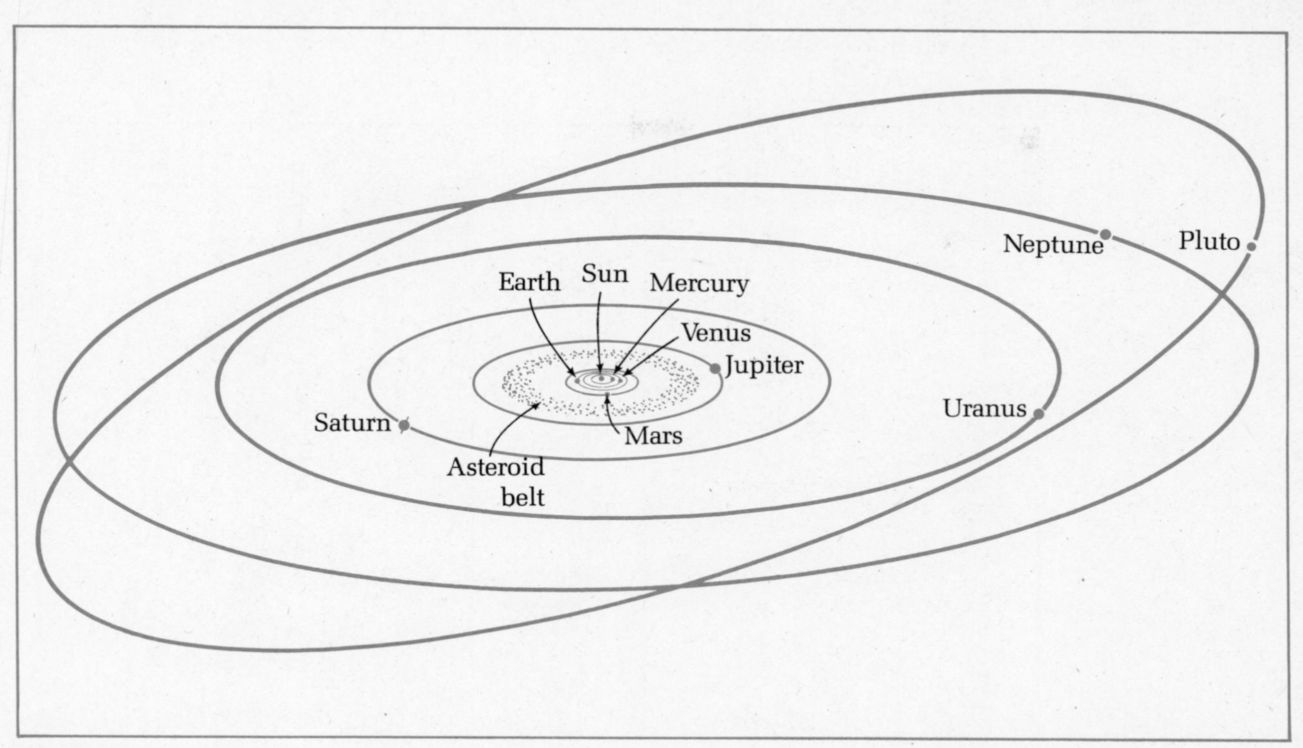

FIGURE 1

As a planet revolves around the sun, it also **rotates,** or spins around like a top. The center of its spin is known as its **axis.** It takes the Earth 24 hours to make one complete rotation around its axis. Earth's axis is tilted in relation to its position to the sun. Because of this, for half of each year one hemisphere is pointed toward the sun while the other hemisphere is pointed away from the sun. The hemisphere that is pointed toward the sun receives more direct sunlight, and experiences the season of summer. The hemisphere that is pointed away receives less direct sunlight, and experiences winter. When Earth has completed half its journey around the sun, the positions of the hemispheres are reversed, and the seasons change.

Lesson 1 Exercise

1. When it is spring in the Northern Hemisphere, what season is it in the Southern Hemisphere?
2. State whether the following is TRUE or FALSE and explain your answer: Before the invention of the telescope, scientists looked into the sky and saw the stars of many galaxies.
3. What causes the moon to orbit Earth?

Item 4 refers to the following material.

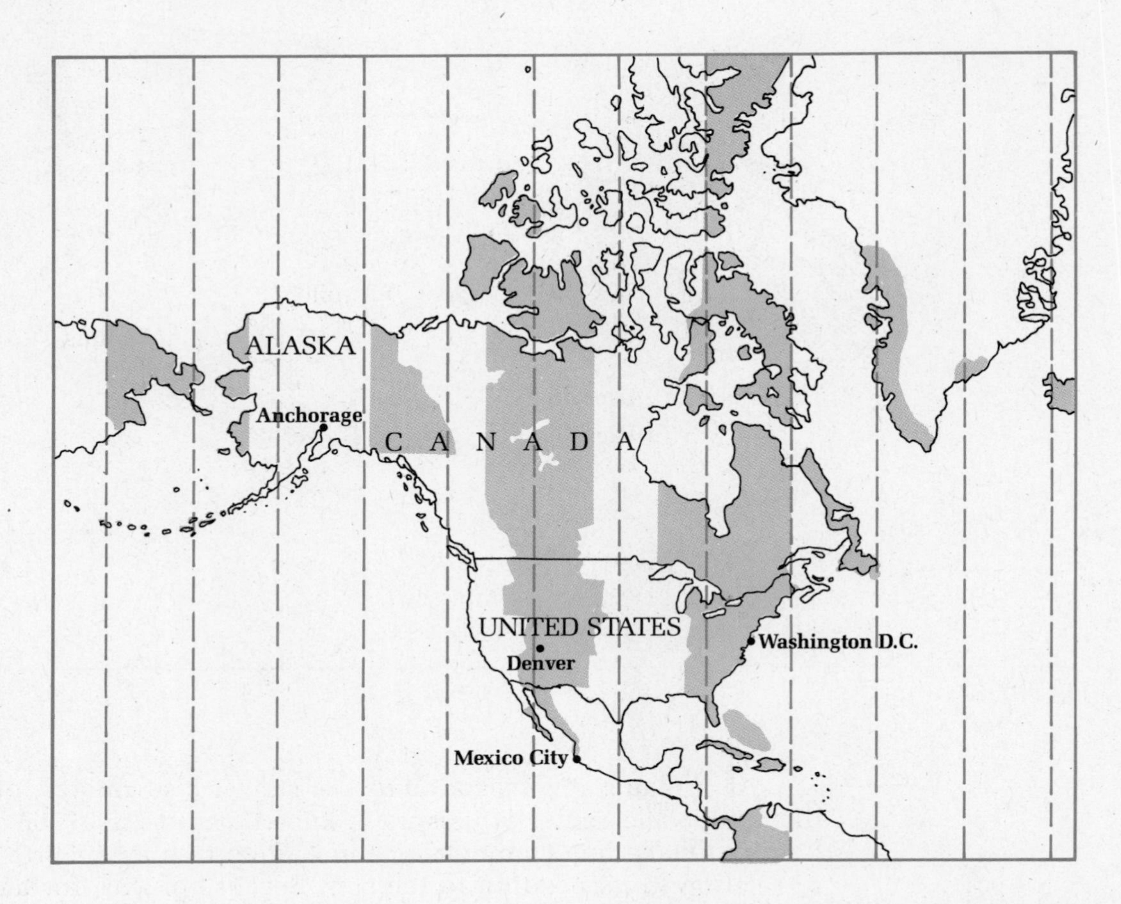

Earth turns in such a way that the sun seems to move west around Earth every 24 hours. The earth is divided into 24 time zones, each about 15° wide, as shown above.

4. You are flying from Washington, D.C., to Denver, Colorado. If you board the airplane at 2:00 P.M. in Washington, D.C., what time should you set your watch for in order to have the correct local time when you land in Denver?

Answers are on page 261.

Lesson 2 History of the Universe and Earth

How was Earth formed and how did it develop?

No one can say for certain how the universe began. A popular theory, however, suggests that it began in a "big bang" some 15 billion to 20 billion years ago. The **big bang theory** states that the universe was an incredibly dense clump of matter which then exploded and expanded outward. The matter was at first completely disorganized. But over billions of years the force of gravity caused particles to cling to one another in larger and larger masses that eventually formed galaxies. The same process occurred within the galaxies as clouds of matter known as **nebulae** gradually condensed into bodies that formed the stars and planets.

It is thought that Earth was formed in this manner about 4.6 billion years ago. At first it was a ball of molten metals and gases. But as it spun on its axis the surface of the planet began to cool, forming a crust. The heavier, still molten metals remained in the center, while the lighter gases formed an envelope around the planet, our atmosphere. In part, this theory is believed to be true because we can still see evidence of this process on Earth today.

The time that Earth has been in existence is called **geologic time.** In order to describe the many ways in which Earth has changed, scientists divide geologic time into units. The larger unit of geologic time is an **era.** Eras are divided into **periods,** and periods are further divided into **epochs.**

There are four eras of geologic time. The first four billion years of Earth's existence make up the **Precambrian Era,** which ended with the emergence of the earliest life forms, mostly a type of bacteria. The next era is the **Paleozoic,** lasting some 345 million years and ending with the emergence of the dinosaurs. The third era, the **Mesozoic,** lasted about 160 million years and ended with the disappearance of the dinosaurs and the emergence of warmblooded life forms. The current era, the **Cenozoic,** has lasted about 65 million years. Humanity has been on Earth only since the very last epoch of the Cenozoic Era, a very small fraction of Earth's history.

Lesson 2 Exercise

1. Scientific dating techniques cannot reveal the age of molten rocks. They can only reveal when the molten rocks cooled and solidified. The oldest rocks that have been found are 4.1 billion to 4.2 billion years old. How is the fact that no older rocks have been found evidence that the first rocks were molten?

2. Based on the description of Earth's formation, which of the following substances would you be MOST likely to find in Earth's core?

(1) oxygen
(2) ice
(3) rock
(4) iron
(5) steam

3. A fossil is any hardened remains of a plant or animal of a previous geological age, preserved in Earth's crust. Bones, footprints, and arrowheads are all examples of fossils. Based on this information and that in the lesson, why do you think the fossil record from the Precambrian Era is the least complete of all four?

Answers are on page 261.

Lesson 3 Structure of Earth

What are the structure and materials that form our planet?

Earth has been slowly cooling since it was first formed. As the rocks and metals within the planet cooled, they formed into layers. The four layers of Earth, shown in the diagram below, are distinguished from each other by their temperature and in part by the type of materials they contain.

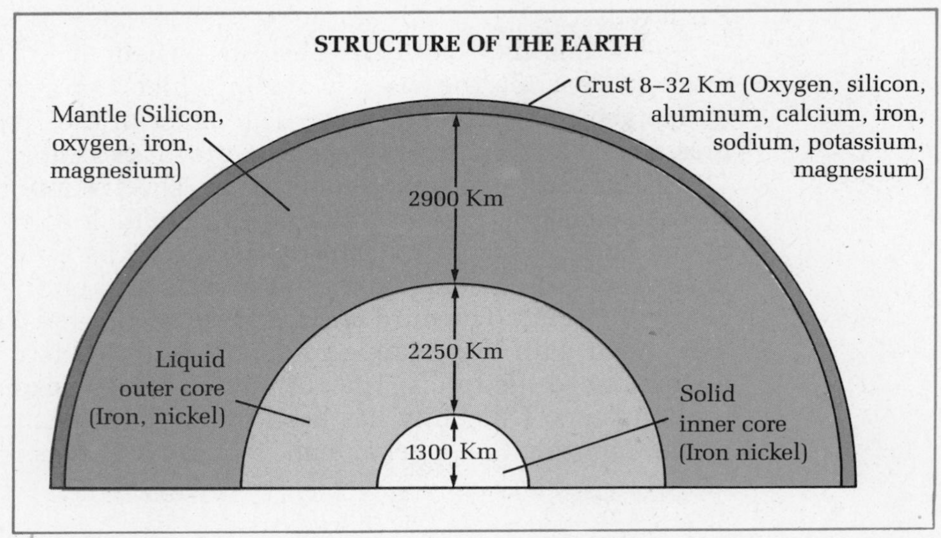

FIGURE 2

Pressure and temperature increase dramatically as one moves from the crust to the core. For this reason, no human being has ever been beneath the Earth's crust. Most of what scientists know about the structure of inner Earth has come from the study of seismic waves. **Seismic waves** are the shock waves that travel through Earth during an earthquake.

We know most about the crust because it is on the crust that we and all living things exist. The rocks in the crust were originally in a molten state known as magma. But various influences in and above the crust can alter magma. Such changes have brought about the three main types of rock that make up Earth's crust.

Magma that is pushed up to Earth's surface, as in a volcanic eruption, is called lava. When lava cools, it forms **igneous rock.** Igneous rock is most often found on or near areas of volcanic activity such as Hawaii. Igneous rocks, such as basalt, are the youngest type of rocks found on Earth's surface. This is because as igneous rocks age, they are subjected to influences which can change them into other forms.

One change rock can go through is to be broken down by wind and water into small particles. When these particles are deposited underwater, they form layers called sediments. The sediments may contain soil, sand, clay, mud, and the remains of living things. In time, the sediments are cemented together by pressure to form **sedimentary rock.** It is in sedimentary rock that evidence of ancient organisms is preserved in the form of fossils. Sedimentary rock, such as sandstone, is the most common type found on Earth's surface. The "stripes" seen on the walls of the Grand Canyon are ancient sediments that have turned to stone.

Rocks that are found deep within the crust may also be altered by its great heat and pressure. The new rock that is formed is called **metamorphic rock.** The veins in marble, which is a type of metamorphic rock, show that it has been "squeezed" into its current form by very high pressure.

Lesson 3 Exercise

1. Indirect evidence is that which is obtained without ever seeing or touching the object being studied. State whether the following statement is TRUE or FALSE and explain your answer: All that scientists know about the interior of Earth has come from indirect evidence.
2. Explain how sandstone could be changed into an igneous rock.
3. Which of the following processes would cause a *sudden* change in Earth's surface?
   (1) sediment deposition
   (2) volcanic eruption
   (3) metamorphic rock formation
   (4) ageing of igneous rocks
   (5) cooling of magma

Answers are on pages 261–262.

Lesson 4 The Changing Earth

What are the forces that cause Earth to change?

Because Earth is still in the process of formation, it is constantly changing. There are many influences that continually alter the planet's surface and crust. Some of these influences come from beneath the surface, pushing out against the crust. Some of them, such as the weight of mountains and oceans, and the action of weather, exist on the surface.

One important subsurface change is caused by movements of plates in Earth's outer shell. The ocean floor and continents are embedded in the plates and share their motion. Some plates move away from each other, allowing new material to push up into the gap between them. Other plates move toward each other. When two plates collide, they can cause folds in the crust, which is how some mountain ranges are formed. It is believed that millions of years ago, all the land on Earth may have been one huge continent that eventually was pulled apart by the motion of the plates into the continents of today.

The point where two plates meet is where most geological activity takes place. **Volcanoes** are often found along these points, as magma beneath Earth's surface rises through cracks and weak spots. Lava flows can greatly alter Earth's surface as they create mountains on land and islands in the oceans. Iceland and Hawaii are islands that are still growing as lava pours out of their volcanoes.

In part, lava is forced from the planet's interior by internal pressure. Strong internal pressure can also cause the crust to fracture, resulting in an **earthquake.** Earthquakes occur deep within the crust, but their effects are felt on the surface. When the crust fractures, cracks called **faults** are created. Pressure also causes rocks along pre-existing faults to move. In both cases, the ground "snaps back" and vibrates as it settles into its new position.

Two examples of surface changes are weathering and erosion. Both of these processes are slow changes. **Weathering** is the breakup of rocks caused by changing temperatures, freezing water, and plant growth. Weathering is the way in which rocks are broken into small enough particles to form soil. **Erosion** is the process by which rocks and soil are worn down and carried away. Wind, rain, ocean waves, rivers, and streams are the main causes of erosion. **Glaciers,** which are large, slow-moving sheets of ice, also cut out land and deposit it elsewhere.

Lesson 4 Exercise

1. Which of the following pieces of evidence would support the idea that plates exist under the ocean floor?

(1) Mountain ranges have been discovered on the ocean floor.
(2) The oceans appear to be getting deeper.
(3) Earth's crust is thinner under the oceans than under the continents.
(4) Earthquakes can cause huge ocean waves.
(5) The floor of the ocean is remarkably smooth and level.

2. How can a subsurface change also cause a surface change?

3. Describe a process that might cause mountain ranges to become smaller.

4. How does planting a winter ground cover, such as alfalfa, decrease erosion of a field?

Answers are on page 262.

Lesson 5 Earth's Water

How does water affect life on Earth?

Earth has often been referred to as the "water planet" because so much of its surface is covered by water. Seventy percent of Earth's surface is water, and 97 percent of that water is in the oceans.

Ocean water is called salt water because it contains salt, as well as many other dissolved minerals. Inland rivers and streams, on the other hand, contain no salt, and because of this their water is called fresh water. Humans and other land-living organisms could not live without fresh water. Other sources of fresh water include lakes, and water beneath Earth's surface called groundwater, which supplies water for our wells. Considerable amounts of fresh water are also frozen in polar icecaps and glaciers.

The supply of fresh water on Earth is constantly being renewed through a process known as the **water cycle.** In the water cycle, heat from the sun causes water from the oceans, rivers, and lakes to turn into vapor, which then rises into the atmosphere during **evaporation.** As the water vapor rises, it cools, returning to its liquid form as droplets that form clouds during **condensation.** Eventually, some of the droplets become heavy enough to fall to Earth as **precipitation,** in the form of rain, snow, sleet, or hail.

Precipitation is fresh water because the ocean salts were left behind when the water evaporated. After the water falls to Earth, some of it evaporates to begin the cycle over again. (See Figure 3.)

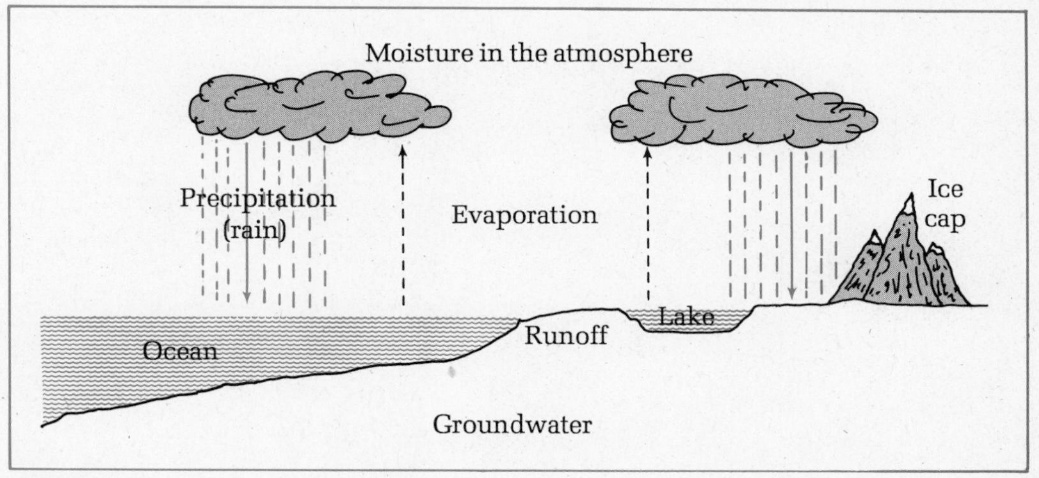

FIGURE 3

Lesson 5 Exercise

1. What are the three steps in the water cycle? Why do you think this process is called a cycle?

2. Some scientists believe that pollution is raising the temperature of Earth's atmosphere. As a result, they fear that the polar icecaps might melt. Which of the following statements is NOT a possible consequence of the melting of the polar icecaps?
   (1) The level of Earth's oceans would rise several feet.
   (2) Ocean water would be diluted, causing many types of fish that depend on salt water to die.
   (3) Because salt retains heat, the general temperature of the oceans will drop.
   (4) Coastal cities would be flooded.
   (5) Fresh water would become scarce.

Items 3 and 4 refer to the following passage and illustration.

Tides, periodic changes in the elevation of the ocean surface at a particular place, result from the gravitational attraction of the moon, and to a lesser extent, the sun. When a certain part of the ocean faces the moon, the water is heaped up toward it by the gravitational pull. At the same time, on the opposite side of Earth, another high tide is created. This is because the solid Earth is pulled more toward the moon than the water, since Earth is closer to the moon on that side. The water is left behind as Earth pulls away, creating a second high tide.

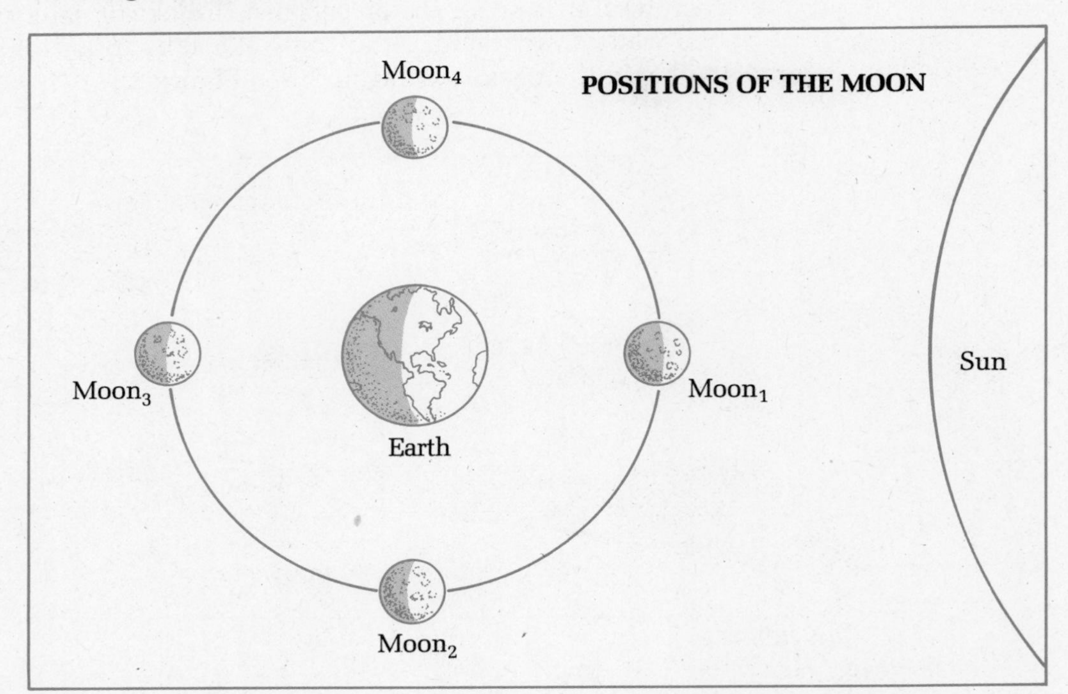

3. Which two positions of the moon produce the highest tides?

4. Moonrise occurs fifty minutes later each day, affecting the tides accordingly. Yesterday high tide was 7:50 A.M. If you wanted to set sail at high tide this morning, around what time would you leave?

Answers are on page 262.

Lesson 6 The Atmosphere

Why is the atmosphere necessary for life on Earth?

The atmosphere is the envelope of air that surrounds Earth. It is about 350 miles thick and is held to the planet by gravity. The atmosphere consists mainly of nitrogen, a gas that is harmless to animals but is used by plants to make protein. The oxygen we breathe makes up just over one-fifth of the atmosphere. The atmosphere also includes carbon dioxide and other gases, such as argon, the gas used in light bulbs. Besides gases, the atmosphere also contains water vapor and large quantities of dust.

The atmosphere is divided into four layers on the basis of temperature (see Figure 4). The **troposphere** touches Earth and is the layer in which we live. Almost all weather takes place within the troposphere. Though the troposphere can be as much as 11 miles thick, most of the air it contains is within 3 miles of the planet's surface. This is why mountain climbers sometimes need oxygen masks at high altitudes.

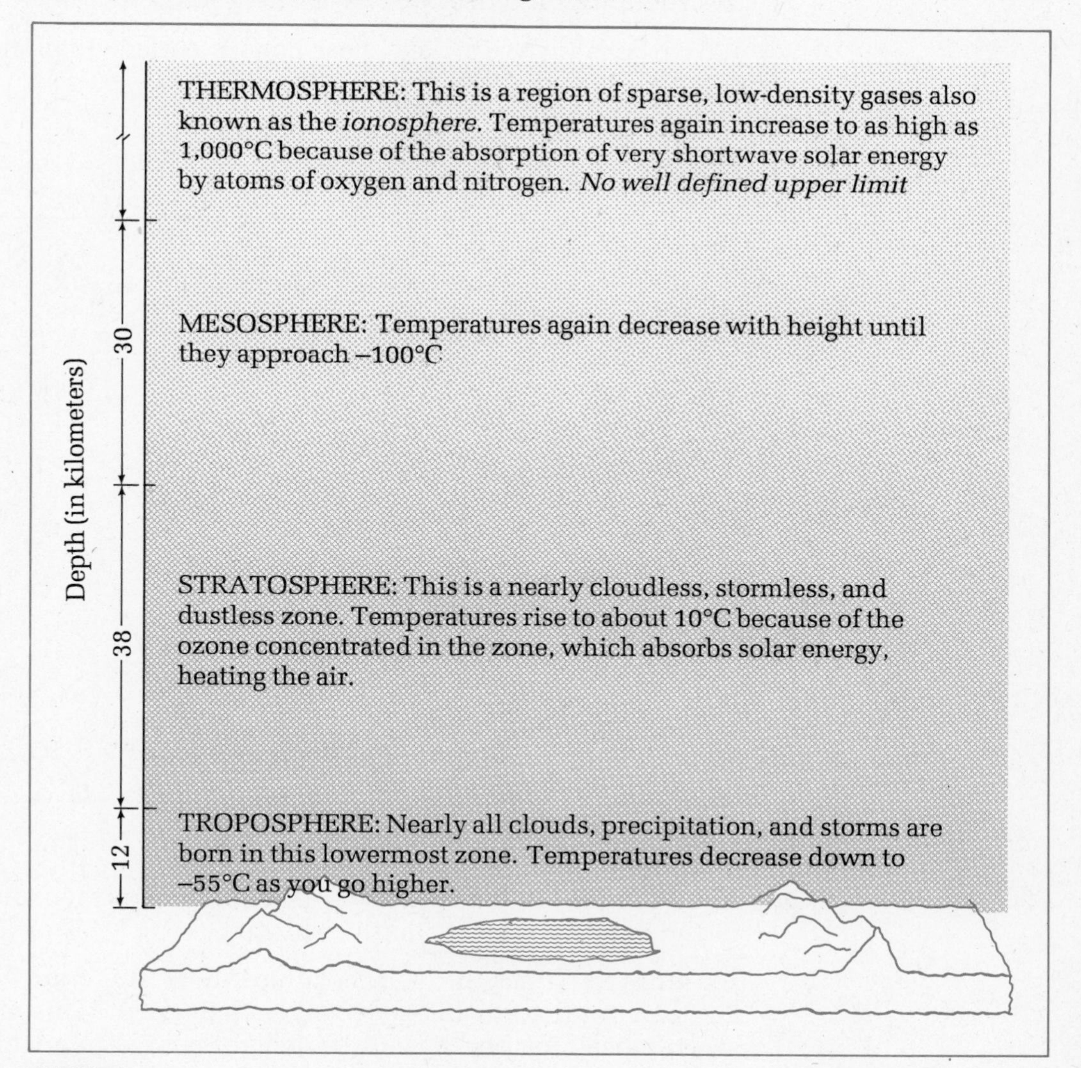

FIGURE 4

Like water, the atmosphere is, in part, responsible for shaping the planet and life on it. To carry out life processes, living things require the oxygen, nitrogen, and carbon dioxide found in the atmosphere. The winds carry heat and water around the globe, and contribute to rock erosion and, thus, to soil formation. The atmosphere also protects Earth from harmful elements from outer space. Particles in the upper atmosphere cause meteoroids (large chunks of matter from outer space) to burn up before they hit the ground. The **stratosphere** contains a layer of **ozone,** a form of oxygen gas which filters out harmful ultraviolet radiation from the sun. Ultraviolet radiation in small doses causes us to tan, but in large doses it can burn, cause skin cancer, and even blind people.

Lesson 6 Exercise

1. Scientists believe that chemicals used in refrigerants, insulation, foam packaging, and aerosols can destroy atmospheric ozone. Why is the destruction of the ozone layer a threat to human health?
2. Why is air much "thicker" near Earth's surface than at high altitudes?

Items 3 and 4 refer to the following figure.

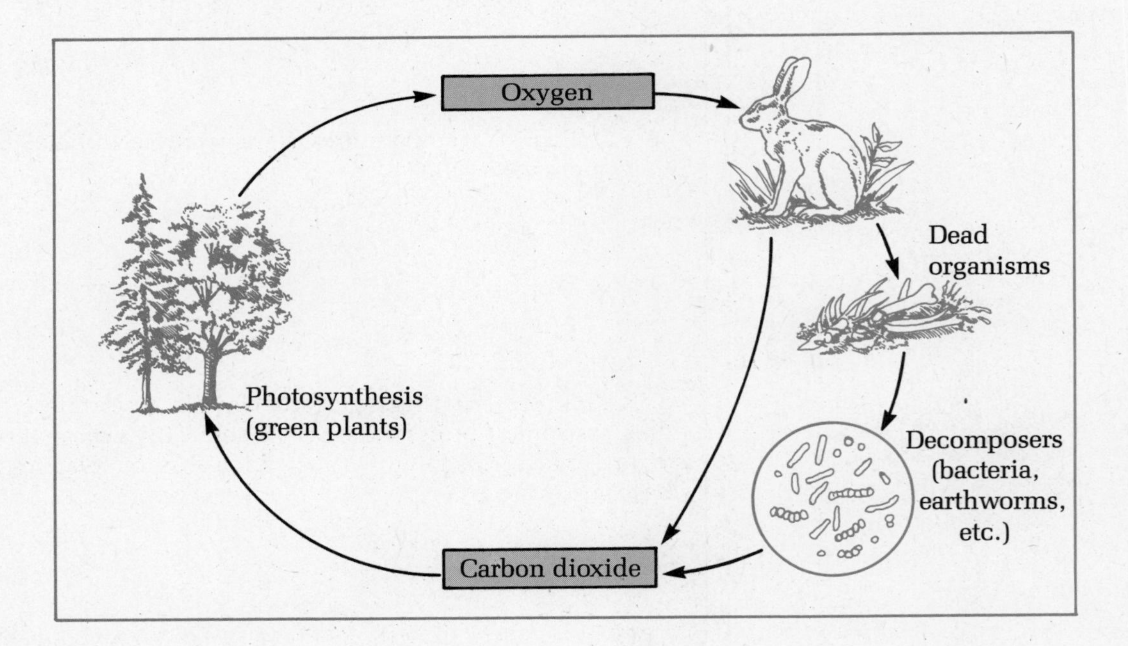

3. According to the figure, how is the supply of oxygen and carbon dioxide in the atmosphere constantly renewed?
4. As sunlight passes through the atmosphere, it is trapped by water vapor and carbon dioxide and reflected back to Earth. This effect is called the **greenhouse effect** because it traps light energy in a way similar to glass

in a greenhouse. The greenhouse effect keeps Earth warm, even at night, acting as a sort of a blanket to retain the sun's heat. Using this information and that in the figure, explain how cutting down vast forests could lead to a warming of the atmosphere.

Answers are on page 262.

Lesson 7 Weather and Climate

What are the main causes of weather and climate?

Like everything else on Earth, the atmosphere is in a constant state of change. Changing conditions in the atmosphere are seen in the weather and climate. **Weather** is the state of the atmosphere in a particular place and time. **Climate** is a summary of the weather conditions in a particular place over a long period of time.

Climate is affected by a region's altitude, or its height above sea level, and by its proximity to large bodies of water. For example, the oceans absorb solar heat and move it from the equator to other regions. Climate is also affected by a region's location on the globe: the closer a region is to the equator, the hotter its climate is likely to be.

Weather is affected by many things, but the most important factor is the sun. It is the sun's energy that warms the air, evaporates water, and helps to cause winds that move the weather across Earth's surface.

As air is heated by the sun, it rises high into the atmosphere. At the same time, cooler, heavier air is falling. As they exchange places, they set up air currents, called **convection currents,** which create winds. Figure 5 illustrates how winds are created by unequal heating in the atmosphere.

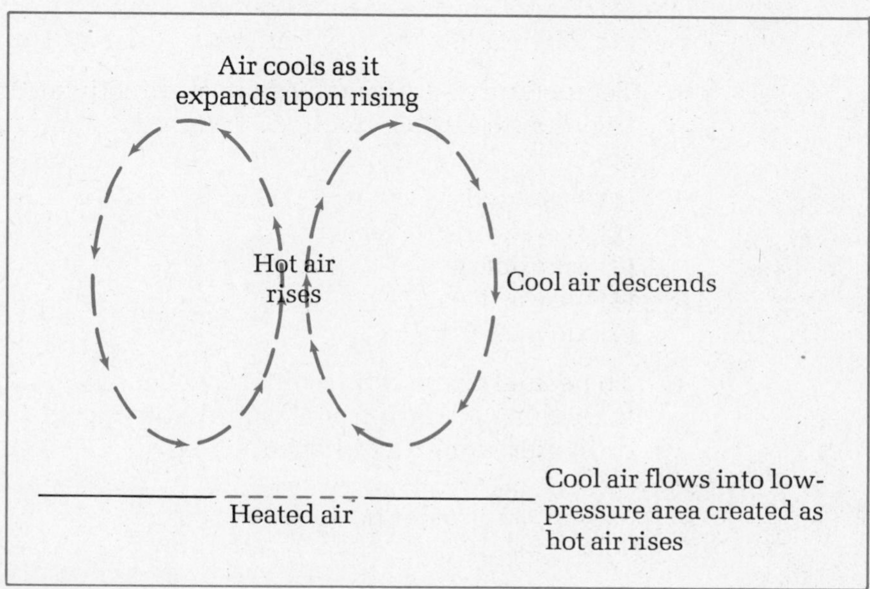

FIGURE 5

Winds are also formed by changes in air pressure. Because warm air has a lower pressure than cool air, it is constantly being pushed aside by the heavier, cooler air. Winds are created by air moving in this way from high-pressure areas to low-pressure areas. Earth's rotation is a third influence on the winds. A complex interaction between the rotation and convection currents sets up wind patterns across the entire globe, called **prevailing winds.** In the areas around the poles and the equator, the prevailing winds move from east to west. In the regions between the equator and the poles, the winds usually blow from west to east in both hemispheres.

Air that is heated to the same temperature tends to form pockets, called **air masses,** that float low in the troposphere. They are pushed along by the winds, and often collide with each other. When two air masses of different temperatures meet, the line between them is called a **front.** As fronts form and move, a change of weather usually takes place. A **warm front** occurs when warm air moves in to replace cold air; a **cold front** occurs when cold air replaces warm air. Weather tends to be cloudy and rainy along either type of front.

Lesson 7 Exercise

1. What is wrong with the following weather forecast? "Tonight, a large warm air mass will move into our region and collide with a cold air mass. We can expect our weather to be sunny and mild for the next several days."

2. Based on information in the lesson, which of the following statements is true?

(1) Warm air masses are usually larger than cold air masses.
(2) High pressure causes stronger winds than low pressure.
(3) A cold front is more likely to affect the weather than a warm front.
(4) Warm air moves more rapidly than cold air.
(5) A warm air mass is less dense than a cold air mass.

3. Temperatures are usually hottest directly at the equator. Before deflection due to Earth's rotation, this causes the principal direction of air to be

(1) eastward
(2) westward
(3) stationary
(4) upward
(5) downward

4. The actual temperature difference along a cold front tends to be greater than along a warm front. Would you expect precipitation to be heavier or lighter along a cold front?

Answers are on pages 262–263.

Lesson 8 Energy and the Environment

What sources of energy help to shape this planet?

We have seen the many ways in which Earth is constantly changing. In order for these changes to come about, they require energy to power them. What are the sources of this natural energy?

By far the most important source of energy is the sun. Energy comes to Earth directly from the sun in the form of light and heat. Without these, Earth might be a frozen lump of rock surrounded by ice, like other planets further from the sun. Sunlight melts Earth's ice, releasing the water which helps to shape the environment and give life to all living things. It heats the air and evaporates water to create weather. The chemical energy in food and wood is available because plants convert sunlight into food energy during photosynthesis.

Directly or indirectly, most of the energy that humans use comes from the sun. The wood we burn exists because of the sun. Likewise, fossil fuels such as coal, oil, and natural gas were created by plants and animals that lived and stored energy from the sun millions of years ago. We produce electricity by burning these fuels, or by harnessing the energy of flowing water in dams and the energy of wind in windmills. We can also make electricity using solar energy which converts sunlight directly.

Another type of energy is geothermal energy, the heat and pressure created deep within the planet. It is geothermal energy that is responsible for plate movement, earthquakes, and volcanoes. Geothermal energy is capable of being harnessed in those areas where underground heat is great enough and close enough to the surface to cause geysers and hot springs. Water that is heated by geothermal energy can be used to generate electric power and to heat homes.

The newest source of energy available to humans is nuclear energy, which is the energy stored in atoms. During nuclear fission, an atom is broken apart, setting off a chain reaction that releases energy as heat. This heat is then used to generate electricity.

Although there are many sources of energy, the main source of energy in the United States today is oil, a fossil fuel. At the current rate of consumption, fossil fuels will probably run out within the next few hundred years. It is possible that fossil fuels can be replaced by solar, nuclear, geothermal and other sources of energy. However, these energy sources have yet to be developed to the point where they can meet the demands of modern society. In addition, nuclear energy poses the problem of producing radioactive wastes, which are difficult to dispose of in a safe manner.

Lesson 8 Exercise

1. A renewable resource can be replaced after it has been used. A nonrenewable resource cannot be replaced once it has been used. Classify each of the following resources as renewable or nonrenewable: geothermal energy, oil, solar energy, wind energy, coal, flowing water.

2. Which of the following statements is NOT supported by the lesson?
   (1) Earth has certain sources of energy that other planets do not have.
   (2) It is important to develop alternative sources of energy.
   (3) When fossil fuels run out, it will not be possible to generate electricity.
   (4) The energy in food comes ultimately from the sun.
   (5) Plants are an important source of energy on Earth.

3. The source of heat for most hot springs is cooling igneous rock. Which of the following statements best explains why over 95% of the hot springs in the United States are found in the West?
   (1) Groundwater is closest to the surface in the West and thus is heated better by the sun.
   (2) Earthquakes are more prevalent in the West.
   (3) Volcanic activity is most recent in the West.
   (4) Groundwater is well insulated by rocks in the West and thus retains heat better.
   (5) The West contains large deposits of fossil fuels.

Answers are on page 263.

Chapter 3 Quiz

1. What causes winter to occur in the Northern hemisphere?

2. Describe the difference between igneous rocks and metamorphic rocks.

3. Explain how unusually warm weather could affect the water cycle in a particular area.

4. Which of the following are examples of weather? of climate?
   (a) a yearly rainfall of 15.5 inches
   (b) temperatures below normal during the first week of April
   (c) a storm caused by cold air from the North meeting warm air from the South
   (d) a trend toward cooler summer temperatures the last fifty years

Items 5 to 7 refer to the following figure.

THE SOLAR SYSTEM

Planet	Average Distance from Sun (kilometers)	Temperature Extremes (Degrees Celsius) High	Low	Major Components of Atmosphere
Mercury	58,000,000	427	–170	none
Venus	108,000,000	480	–33	Carbon dioxide
Earth	150,000,000	58	–88	Nitrogen, oxygen, water vapor, carbon dioxide
Mars	228,000,000	–31	–130	Carbon dioxide, nitrogen, argon, oxygen
Jupiter	778,000,000	29,700	–95	Hydrogen, helium, methane, ammonia
Saturn	1,427,000,000	?	–180	Hydrogen, helium, methane, ammonia
Uranus	2,869,000,000	?	–210	Hydrogen, helium, methane
Neptune	4,486,000,000	?	–220	Hydrogen, helium, methane
Pluto	5,890,000,000	?	–230	Methane

5. On Earth, carbon dioxide absorbs and retains the sun's heat after it bounces off the ground back into the atmosphere through a process called the greenhouse effect. Which of the following statements is evidence that the greenhouse effect operates on Venus as well?

A. Venus has carbon dioxide in its atmosphere.
B. Venus has higher average temperatures than Mercury does.
C. Venus is closer to the sun than Earth is.

(1) A only
(2) B only
(3) C only
(4) A and B only
(5) A and C only

6. Which of the following statements supports the possibility that Mars could contain life?

(1) Mars and Earth both have solid crusts.
(2) Animals on Earth take in oxygen and give off carbon dioxide.
(3) Temperatures on Mars never go above the freezing point of water (0°C).
(4) Mars is closer to the sun than five other planets.
(5) Plants and animals on Earth respire water vapor.

7. Assuming ammonia is present on Uranus and Neptune, why isn't it found in the atmosphere, as on Saturn and Jupiter?

---

8. If an artificial satellite were to gain enough speed to escape the force of Earth's gravity, what would be its path of travel? Where would it land?

9. San Francisco is due south of Seattle, yet if an airplane pilot were to set a course due south of Seattle, the airplane would end up somewhere west of San Francisco over the Pacific Ocean. Which of the following statements best explains why this would happen?

   (1) Earth rotates eastward under the plane as it flies south.
   (2) Earth rotates westward under the plane as it flies south.
   (3) Convection currents in the atmosphere push the plane westward.
   (4) Wind resistance causes the plane to drift westward.
   (5) Earth's gravity pulls the plane westward.

10. Friction between Earth and the ocean waters causes a drag on Earth as it rotates. The effect of the friction is to slow Earth's rotation. Would this cause a day to get gradually longer or shorter?

Answers are on page 263.

Chapter

4 Chemistry

Objective

In this chapter you will read and answer questions about

- Matter and its properties
- Atoms and molecules and how they combine
- Elements and compounds
- Chemical reactions and equations
- Energy changes in chemical reactions
- Radioactivity and nuclear energy

Lesson 1 Properties of Matter

What is matter? What are its properties?

Matter is anything that has an amount of material, or **mass,** and takes up space, or has **volume.** We can usually see, touch, or smell matter, but even those things that we cannot perceive—such as certain gases—are made of matter. Water, air, wood, silver, plastic, helium, and glass are all examples of matter.

There are three different phases, or **states,** in which matter can exist on Earth. These states are solid, liquid, and gas. You can identify a **solid** because it has definite shape, mass, and volume. This book is an example of a solid. A **liquid** has definite mass and volume, but no definite shape. When you pour water into a glass, it will take the shape of the glass. A **gas** also has definite mass, but it does not have a definite shape or a definite volume. A gas will spread out to fill the size and the shape of its container.

Matter can change from one state to another. Such changes include melting, freezing, evaporation, boiling, and condensation.

Different types of matter can be identified by different characteristics, or properties. Properties that you can observe with your senses are called **physical properties.** These include size, shape, color, mass, texture, volume, taste, and smell. To describe a lump of sugar as being white, sweet-tasting, and cube-shaped is to describe it in terms of physical properties.

Matter can also be described in terms of **chemical properties.** Chemical properties describe how a substance will combine with other substances. When wood burns, it combines with oxygen to form a new substance, ash.

A colorless, odorless gas is also produced. Wood's ability to combine with oxygen is an example of a chemical property.

When only the physical properties of a substance change, the change is called a **physical change.** In physical change, the identity of the substance stays the same. Chopping a piece of wood into smaller pieces is an example of physical change, because even though the wood looks different, it is still wood. When a substance changes from one state to another, the change is also physical. Water is still water whether it is in its liquid state or whether it is in the form of ice or steam.

When the identity of a substance changes, the change is called a **chemical change.** Chemical changes produce new kinds of matter. The rusting of iron is a chemical change because the iron combines with oxygen to form a new substance called iron oxide, or rust.

Lesson 1 Exercise

1. When frost melts, it turns directly into a gas without first becoming a liquid. Such a change is called sublimation. Is sublimation a physical or a chemical change? Explain your answer.
2. Explain why fuel efficiency is increased by using a gas rather than a liquid to inflate car tires.
3. Which of the following statements is evidence that the changes leaves undergo in the fall are chemical changes?
   (1) The leaves fall off the trees.
   (2) The leaves turn colors.
   (3) The leaves can no longer make food for the trees.
   (4) The size and shape of the leaves do not change.
   (5) Shorter days and lower temperatures trigger the changes.

Answers are on page 263.

Lesson 2 Structure of Matter

What is the basic unit of matter? How does it bond to form larger units?

If you were to take any substance and try to break it into smaller parts, you would eventually reach the smallest particle. You could not divide this particle without destroying the substance. The smallest particle of a substance that still retains the properties of that substance is called a **molecule.** If you could divide a molecule still further, you would find atoms. **Atoms** are basic units of matter. Atoms may or may not have the same properties as the molecule they are a part of. For example, the individual atoms that make up table sugar do not taste sweet. Different kinds of matter differ in the kinds of atoms present and how the atoms are arranged.

Scientists have discovered that an atom consists of particles located in a central core, or **nucleus,** and particles whirling around the outside of the nucleus. Some atomic particles have electrical charges, although atoms themselves are usually neutral. That is, whole atoms have no electrical charge as such. How can this be?

Electrical charges can be positive or negative. A positive electrical charge balances one that is negative, making the total charge neutral. For this reason, we can see that within an atom the positive and negative charges combine to cancel each other out.

The particle in an atom that has a positive charge is called a **proton.** Protons are found in the nucleus of the atom. Another particle that is found in the nucleus is the **neutron,** which has no electrical charge, and therefore does not affect the electrical balance of the atom.

The particle that provides the negative charge in an atom is called the **electron.** In a neutral atom, the number of electrons always equals the number of protons. Electrons are located in the area surrounding the nucleus. At one time it was believed that electrons circled the nucleus in fixed orbits the way that planets orbited the sun. Scientists know today, however, that this is not the case. Electrons occupy a general region around the nucleus that is more like a cloud than an orbit. An **electron cloud** refers to the area in an atom where an electron is most likely to be found.

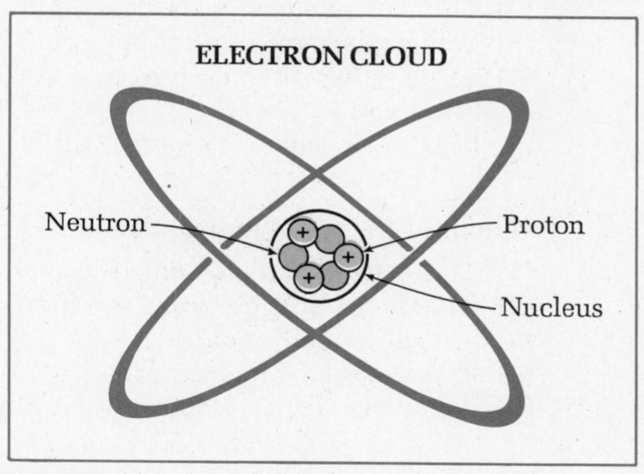

FIGURE 1

Electrons exist in different energy levels. Those electrons closest to the nucleus are in the first, or lowest, energy level. Those electrons furthest from the nucleus are in the highest energy level and are able to separate from the atom and join other atoms. When this happens, the atom that has lost an electron is no longer neutral. It becomes positively charged because it has one more proton than electron. The atom that gains an electron becomes negatively charged. Atoms that have positive or negative electrical charges are called **ions.**

Ions have the ability to join, or **bond,** with each other. They do this by transferring electrons between them. An ion that is negatively charged can "lend" its spare electron to an ion that is positively charged. Their opposite charges keep them together. Other atoms bond by sharing electrons equally between them. This is called **covalent bonding.** Covalent bonds form molecules. Molecules can be made up of two identical atoms, or made up of two or more different atoms. The bonding of atoms forms the basis of all chemical reactions and combinations.

Lesson 2 Exercise

1. If you break apart hydrogen peroxide, a common household chemical, you end up with two hydrogen atoms and two oxygen atoms. Separately, hydrogen and oxygen have different chemical properties than hydrogen peroxide does. Which units of matter would you classify hydrogen and oxygen as?

2. Table salt is formed when an atom of chlorine that has seven outer electrons captures the single outer electron of the sodium atom. This electron now stays mostly with the chlorine atom. Is the formation of sodium chloride an example of covalent bonding or ionic bonding?

3. Any chemical reaction that involves the loss of an electron by an atom is called oxidation. A chemical reaction that results in a gain of an electron by an atom is called reduction. Which of the following statements proves that the formation of table salt from sodium atoms and chlorine atoms is an oxidation-reduction reaction?

   (1) Chlorine captures the single outer electron of the sodium atom.
   (2) Chlorine is a gas in its pure state.
   (3) Table salt has a neutral electrical charge.
   (4) Table salt separates into sodium atoms and chlorine atoms when dissolved in water.
   (5) Passing an electrical current through salt water makes sodium and chlorine separate faster.

4. Which of the following statements is NOT true of an ion?

   (1) Its electrons form an electron cloud.
   (2) Its protons and neutrons are found in its nucleus.
   (3) It is electrically neutral.
   (4) It can bond with another ion.
   (5) It can transfer an electron to another ion.

Answers are on pages 263–264.

Lesson 3 Elements and Compounds

What does the periodic table tell us about elements?

Substances that are made up of only one type of atom are called **elements.** Elements are substances that cannot be broken down into simpler substances. Silver, oxygen, tin, and hydrogen are some common elements. A substance like water is not an element because it can be broken down into the elements hydrogen and oxygen.

There are 109 known elements, which scientists organize in a chart called the **periodic table.** See the periodic table on pages 128–129. In the periodic table, elements are arranged according to the number of protons that they contain. The number of protons in an atom, known as the atomic number, determines the characteristics of the atom. For example, all atoms that have just one proton are hydrogen atoms, while all atoms that have two protons are helium atoms. Because of this, no two elements have the same atomic number.

As you look at the periodic table, you will notice that each element is listed not by name, but by one or two letters. These abbreviations are the chemical symbols for the elements. The first letter (or in some cases, the only letter) of a chemical symbol is always capitalized.

The periodic table tells us more about an atom than its symbol and atomic number. The elements in the periodic table are also arranged according to their properties. Elements that have similar properties are arranged in vertical columns, called groups. For instance, the group on the far right, which includes the gases helium and neon, contains all the elements that are not able to combine easily with other elements. In addition, all the metals are on the left side of the table; all the nonmetals, except for hydrogen, are on the right side of the table. **Metals** tend to be those elements that are shiny, that can be drawn out into thin wires, and that are good conductors of heat and electricity. **Nonmetals** tend to be poor conductors of heat and electricity.

Not all elements are able to combine with one another, but most are. As we have seen, some elements will give up electrons while other elements will take electrons. Therefore, metals combine best with nonmetals because metals are "lenders" of electrons, while nonmetals are usually "borrowers" of electrons. Whatever the elements involved in such a combination, the chemical bonding of two or more elements always results in a new substance, called a compound.

A **compound** is made up of two or more elements chemically combined. For instance, water is a compound. The molecules in water are made up of one oxygen atom chemically bonded to two hydrogen atoms. The elements hydrogen and oxygen are not altered themselves, but their combination gives the compound its own particular properties.

Just as elements can be written using symbols, compounds can be written using formulas. A chemical formula shows the type and number of atoms in each molecule of a given compound. For example, the formula for water is H_2O. The small number 2, known as the **subscript**, indicates how many atoms of hydrogen are in this compound. Therefore, a molecule of water contains two hydrogen atoms and one oxygen atom.

The Periodic Table

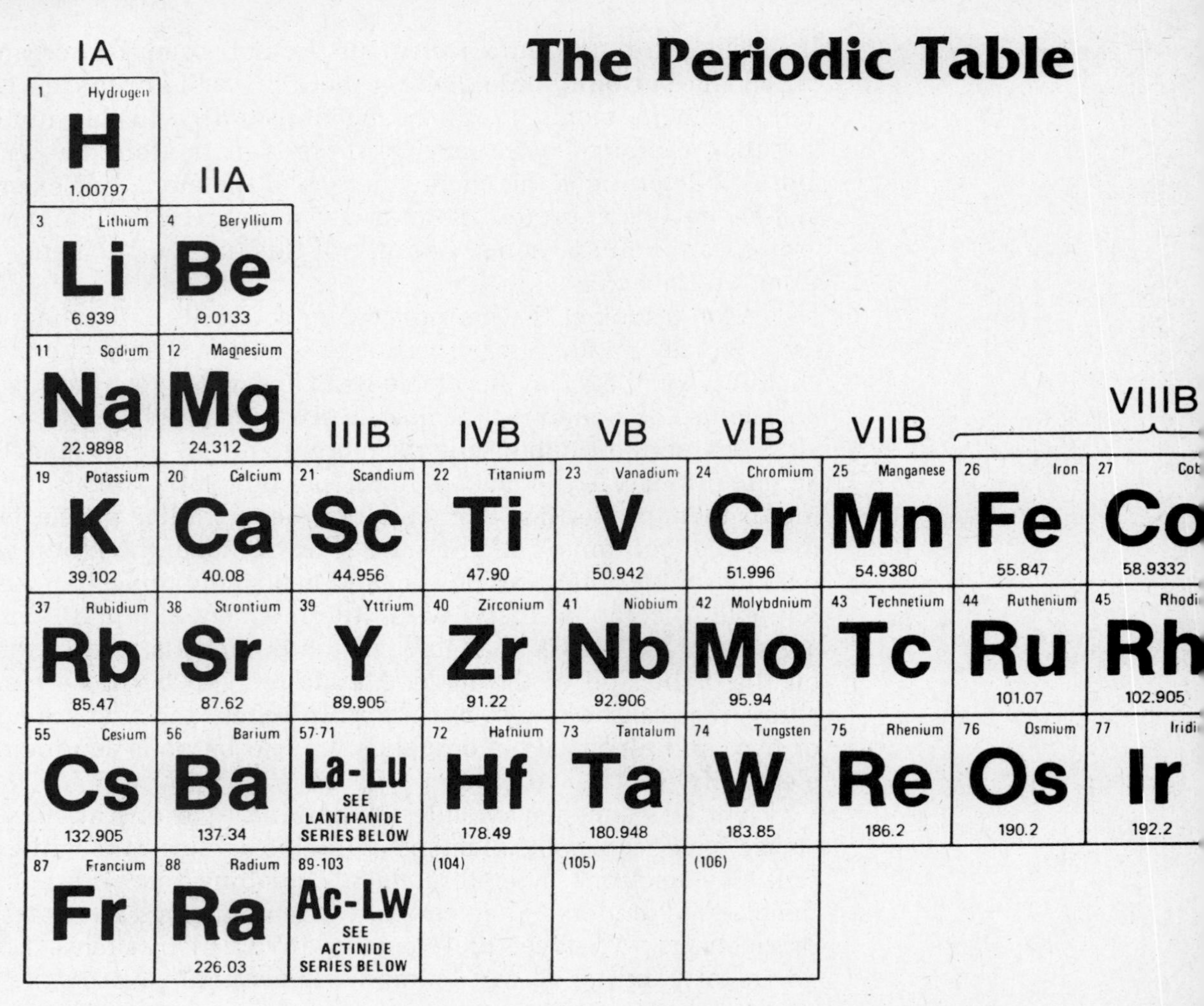

FIGURE 2

Lesson 3 Exercise

1. One molecule of vinegar (acetic acid) contains four atoms of hydrogen, two atoms of carbon, and two atoms of oxygen. Explain why vinegar is not an element.

# of Elements

	IB	IIB	IIIA	IVA	VA	VIA	VIIA	O
								2 Helium **He** 4.0026
			5 Boron **B** 10.811	6 Carbon **C** 12.01115	7 Nitrogen **N** 14.0067	8 Oxygen **O** 15.9994	9 Fluorine **F** 18.9984	10 Neon **Ne** 20.183
			13 Aluminum **Al** 26.9815	14 Silicon **Si** 28.086	15 Phosphorus **P** 30.9738	16 Sulfur **S** 32.064	17 Chlorine **Cl** 35.453	18 Argon **Ar** 39.948
28 Nickel **Ni** 58.71	29 Copper **Cu** 63.54	30 Zinc **Zn** 65.37	31 Gallium **Ga** 69.72	32 Germanium **Ge** 72.59	33 Arsenic **As** 74.9216	34 Selenium **Se** 78.96	35 Bromine **Br** 79.909	36 Krypton **Kr** 83.80
46 Palladium **Pd** 106.4	47 Silver **Ag** 107.870	48 Cadmium **Cd** 112.40	49 Indium **In** 114.82	50 Tin **Sn** 118.69	51 Antimony **Sb** 121.75	52 Tellurium **Te** 127.60	53 Iodine **I** 126.9044	54 Xenon **Xe** 131.30
78 Platinum **Pt** 195.09	79 Gold **Au** 196.967	80 Mercury **Hg** 200.59	81 Thallium **Tl** 204.37	82 Lead **Pb** 207.19	83 Bismuth **Bi** 208.980	84 Polonium **Po**	85 Astatine **At**	86 Radon **Rn**

64 Gadolinium **Gd** 157.25	65 Terbium **Tb** 158.924	66 Dysprosium **Dy** 162.50	67 Holmium **Ho** 164.930	68 Erbium **Er** 167.26	69 Thulium **Tm** 168.934	70 Ytterbium **Yb** 173.04	71 Lutetium **Lu** 174.97
96 Curium **Cm**	97 Berkelium **Bk**	98 Californium **Cf**	99 Einsteinium **E**	100 Fermium **Fm**	101 Mendelevium **Mv**	102 Nobelium **Nb**	(103) Lawrencium **Lw**

2. Which of the following is NOT something that a single water particle and a single hydrogen particle have in common? Explain your answer.
   (1) mass
   (2) proton
   (3) neutron
   (4) atom
   (5) electron

Item 3 refers to the following figure.

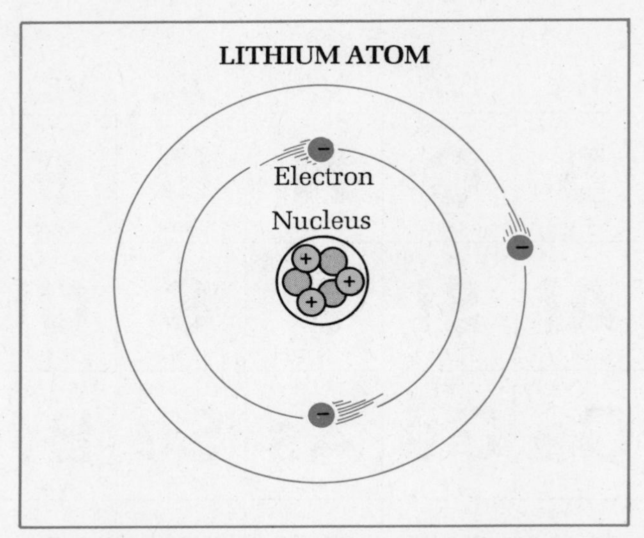

3. The lowest energy level of an atom can contain no more than two electrons. The next energy levels each can contain no more than eight electrons. Compare the diagram of a lithium atom above with the periodic table. Why do you think lithium and hydrogen are in the same group?

---

4. Would you predict the element calcium to be a good conductor of heat and electricity? Why or why not?

Answers are on page 264.

Lesson 4 Mixtures and Solutions

What are the ways that substances can be combined?

Elements and compounds are called **pure substances.** A pure substance contains only one kind of particle, and so has constant properties throughout. Elements contain only one kind of atom, and compounds contain only one kind of molecule.

When two or more pure substances are mixed together, but are not chemically combined, the result is a **mixture.** Because mixtures involve only a physical change and not a chemical change, the substances that are mixed keep their properties. When you mix chocolate with milk, the taste of the milk remains, and you can still see and taste the chocolate in it. Other common mixtures are such things as sand, spaghetti sauce, gasoline, and air.

Sometimes one of the substances in a mixture seems to disappear. When you add sugar to a cup of coffee, the sugar crystals disappear, or **dissolve.** You can still taste the sugar, however, so you know that it has not changed chemically. The sugar crystals have broken up into tiny particles—in this case, sugar molecules. The substance that dissolves, such as sugar, is called a **solute.** The substance in which something dissolves is called a **solvent.** Together, they form a special type of mixture known as a **solution.**

Solutions can involve solids, liquids, and gases, but the most common solutions consist of solids or gases dissolved in liquids. Many common solutions have water as the solvent. Substances that will dissolve in water, such as sugar, salt, and carbon dioxide (to make carbonated water), are said to be **soluble** in water. Not everything is soluble in water, however. For example, oil is **insoluble** in water, which is why another solvent, such as soap, must be used to get oil stains out of clothing.

Sometimes when a substance is dissolved in water the resulting solution is able to conduct electricity. A substance that is able to conduct an electric current when dissolved in water is called an **electrolyte.** Most salts are electrolytes, as are most acids. A car battery is able to produce and conduct an electric current because it contains solutions of electrolytes.

Lesson 4 Exercise

1. Classify each of the following as a solution or as a mixture.

(a) soft drink
(b) salted popcorn
(c) lemonade
(d) sea water

Items 2 to 4 refer to the following passage.

Electrolytes can conduct an electric current because ions exist in solutions of electrolytes. Table salt separates into sodium and chlorine ions when dissolved in water, and so it is capable of conducting electricity when in solution. The aqueous solutions of nonelectrolytes do not conduct electric current. In solution, nonelectrolytes exist as molecules rather than ions. Pure water alone does not conduct electricity.

2. Is a solution of pure water and sugar capable of conducting electricity? Explain your answer.

3. Pure water does not have the ability to conduct electricity, yet tap water conducts electricity, making it dangerous to touch electrical objects while taking a bath or a shower. Why do you think tap water conducts electricity?

4. Which of the following would cause the greatest error in determining that the solution in the following figure contains electrolytes?

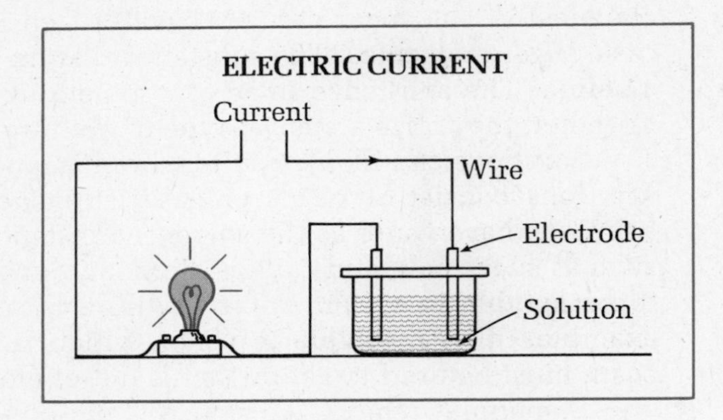

(1) using pure water in the solution
(2) using a metal wire
(3) using a battery as the power source of the electric current
(4) dissolving the test substance in a salty solution

Answers are on page 264.

Lesson 5 Chemical Reactions

How are substances chemically combined with other substances?

Unlike physical changes, chemical changes occur on the molecular level during a **chemical reaction.** A chemical reaction is the process in which electrons are transferred or shared between atoms. In a chemical reaction, molecules can be broken down into their component atoms. If the atoms are electrically charged ions, they may recombine to form new substances. Whatever the reaction, the result will be a substance that has different properties than the original. For instance, hydrogen and oxygen are both colorless gases that burn. But in a chemical reaction, hydrogen and oxygen molecules can break up to form ions, which then combine in a new arrangement to form the compound water, a liquid that has the property of putting out fire.

Chemists write down chemical reactions in a kind of shorthand called a **chemical equation.** A chemical equation is useful because it describes what type of chemical reaction is possible. It also gives the proportions of substances that allow the reaction to occur. For example, the reaction of hydrogen and oxygen to form water is written as follows.

$$2H_2 + O_2 \longrightarrow 2H_2O$$

This equation shows that two molecules of hydrogen and one molecule of oxygen will combine to form two molecules of water. In the equation, the substances at the beginning of the reaction, known as the **reactants,** are written on the left side. The substances that result from the reaction, known as the **products,** are written on the right side. Hydrogen and oxygen are written as H_2 and O_2 because they are normally found in air as molecules consisting of two bonded hydrogen atoms and two bonded oxygen atoms.

In an equation, the number of atoms in the reactants must always equal the number of atoms in the product. You can see that in the equation above, there are four hydrogen atoms and two oxygen atoms on both the right side and the left side of the reaction. This phenomenon is explained by the law of conservation of matter, which states that matter can be neither created nor destroyed. The total number of atoms in a chemical reaction does not change; the atoms are merely rearranged. It is this rearrangement of atoms that gives the products of a reaction different properties from the reactants.

The properties of the products of a reaction are often determined by the proportion of the reactants involved. To make water, hydrogen atoms must combine with oxygen atoms in a 2:1 ratio. If the number of atoms involved in the reaction were changed, the resulting compound would be different. For example, if only two hydrogen atoms bonded with two oxygen atoms, the resulting compound would not be water.

$$H_2 + O_2 \longrightarrow H_2O_2$$

H_2O_2 is the chemical formula for hydrogen peroxide. The difference in the number of atoms in the reactants resulted in a different product.

Chemical reactions occur at different rates. The chemical reaction that causes a flashbulb to flash happens in less than a second, while the chemical reaction that produces rust can take weeks or even months. Some of the factors that affect the speed of a chemical reaction include the type of substances that are involved, the temperature, the amount of each reactant that is present, and whether the reactants are shaken or stirred.

Lesson 5 Exercise

1. Fluoride prevents tooth decay because fluoride ions replace some hydroxide ions in the compound that makes up the tooth enamel. This replacement makes the enamel harder and more resistant to decay. Is the substitution of fluoride for hydroxide a chemical reaction? Explain your answer.

2. Which of the following chemical formulas does NOT represent the molecule of a compound? Explain your answer.

   (1) H_2O
   (2) NaCl
   (3) SO_3
   (4) HCl
   (5) $2O_2$

Items 3 and 4 refer to the following passage.

Gasoline is a mixture of carbon- and hydrogen-containing compounds. At times, the combustion of gasoline by oxygen gas yields carbon monoxide gas (CO) and water vapor. Under other conditions, this reaction yields carbon dioxide gas (CO_2) and water vapor. Carbon monoxide is more harmful to humans than is carbon dioxide.

3. Which chemicals are the reactants in the two reactions described above? Which chemicals are the products?

4. If you were designing a car, which of the following adjustments could you make to decrease the harmful carbon monoxide fumes?

(1) Increase the hydrogen supply to the gasoline.
(2) Increase the oxygen supply to the gasoline.
(3) Increase the carbon supply to the gasoline.
(4) Increase the water supply to the engine.

Answers are on page 264.

Lesson 6 Acids, Bases, and Salts

What are the products of a chemical reaction between an acid and a base?

Acids, bases, and salts are three important groups of chemical compounds that are commonly used in everyday life. Each of these compounds has definite properties of its own, though the elements which make them up may be quite similar. However, the elements in each compound are combined in unique ways.

Acids can be found in orange juice, vinegar, coffee, milk, and carbonated drinks. Car batteries contain acids, as do shampoo and eyewash. An important property of acids is that they all contain the element hydrogen. When an acid is dissolved in water, it produces positive ions of hydrogen ($H+$). Chemists can identify a substance as an acid by testing for the presence of $H+$ ions in solution.

Bases are commonly found in cleaning fluids, such as ammonia. Bases are also present in soaps, and in stomach remedies such as milk of magnesia. All bases contain a negative ion of hydrogen bonded to oxygen ($OH-$), called a hydroxide ion.

Salts are a product of the chemical reaction between acids and bases. The most familiar salt is sodium chloride, or table salt. Many other salts exist in ocean water and within the human body.

Some acids and bases are stronger than others. An acid such as sulfuric acid, which is used in car batteries, is extremely strong, because strong acids conduct electricity better than weak acids. Sulfuric acid causes bad burns if it touches the skin, and it is extremely poisonous. Yet citric acid, a weak acid, is mild enough to be swallowed by people every day as they drink orange juice or eat citrus fruit. Strong bases are highly corrosive and poisonous, but weak bases are commonly swallowed to relieve the symptoms of indigestion.

When an acid and a base are mixed together, a chemical reaction takes place. Because acids and bases are chemical "opposites," they neutralize each other to produce two substances that are neither acid nor base—a salt plus water. Thus the reaction that occurs between an acid and a base is called a **neutralization reaction.** It is this reaction that makes a basic substance such as sodium bicarbonate (baking soda) effective in neutralizing excess acids in the stomach. The general equation for a neutralization reaction is shown below.

$$\text{Acid} + \text{Base} \longrightarrow \text{Salt} + \text{Water}$$

Lesson 6 Exercise

Item 1 refers to the following figure.

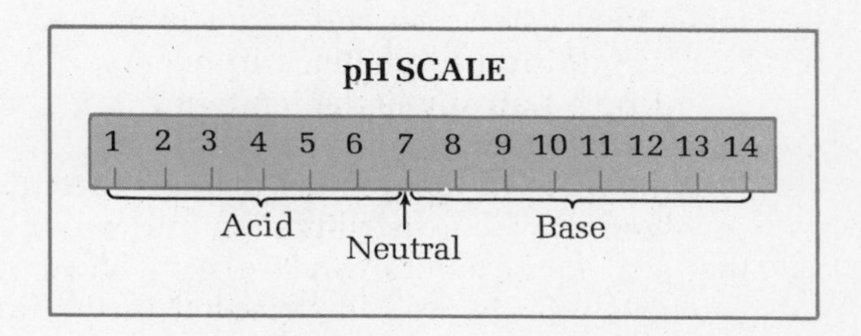

1. Lettuce is a plant that grows best in soils that are slightly alkaline, or basic. Your local county extension office tests the soil in which you wish to grow lettuce and tells you that it has a pH of 5. Which of the following is your best course of action?

(1) Add a substance that will raise the pH of the soil to above 7.
(2) Add a substance that will raise the pH of the soil to 7.
(3) Add a substance that will lower the pH of the soil to below 5.

---

2. A scientist wishes to test the strength of an acid solution. Which of the following tests would be most effective? Explain your answer.

(1) Add water to the solution.
(2) Mix the solution with another acid.
(3) Filter the solution through a solute.
(4) Pass an electric current through the solution.
(5) Taste the solution.

3. Salts that contain sodium are believed to contribute to high blood pressure. Explain why baking soda should not be used as an antacid by people that have high blood pressure.

Answers are on page 264.

Lesson 7 Energy Changes in Chemical Reactions

What are the two ways energy is transferred during a chemical reaction?

All chemical reactions involve changes in energy. In some reactions, such as the rusting of iron, the energy change is hardly noticeable. In other reactions, such as the exploding of a firecracker, the energy change is obvious.

Some chemical reactions give off energy to the environment. This energy may be in the form of heat, light, or an electric current. A reaction in which energy is given off is called an **exothermic reaction.** The burning of wood is an exothermic reaction, as is the flash of a camera flashbulb. A small amount of energy, such as the spark that comes from a match, may be needed to get the reaction started, but the overall effect is a significant release of energy to the environment.

Some chemical reactions absorb more energy than they give off. This type of reaction is called an **endothermic reaction.** Baking a cake is an endothermic reaction, because a considerable amount of heat is required. The process of photosynthesis, in which plants use energy from sunlight to make food, is also an endothermic reaction.

In addition to changes in energy, chemical reactions involve changes in entropy. **Entropy** is the tendency of a system to achieve the most random or disordered arrangement possible. An increase in entropy means that there is an increase in the number of ways that energy is distributed. The particles involved may be distributed more randomly as well. Chemical reactions that are exothermic always involve an increase in entropy, because energy becomes more widely distributed in the universe. Chemical reactions that are endothermic involve a decrease in entropy, because the energy becomes concentrated.

You can better understand the meaning of entropy if you think of a crowded elevator. While the elevator is moving, the people inside are confined to a very small space in a fairly ordered arrangement. Each time the elevator stops, however, people get out until they are distributed all over the building. If you think of people as representing energy, there are now many more ways in which the energy they carry is distributed. Their distribution is also more random; one may be on the second floor, one on the fifth, and so on. The entropy of the system has increased.

Entropy is useful in explaining why reactions occur. Often a reaction will take place because it results in the lowest possible energy state or the greatest possible energy distribution. A reaction may also result in a more random distribution of the particles involved.

Lesson 7 Exercise

1. On which side of a chemical equation would you write the word *energy* for an exothermic reaction? On which side would you write *energy* for an endothermic reaction?

2. Is a melting ice cube undergoing an exothermic or an endothermic reaction?

3. Chemical reactions in which elements combine and form compounds are generally exothermic, while chemical reactions in which stable compounds are broken up into their component elements are usually endothermic. Predict whether or not energy is necessary to produce hydrogen and oxygen from water.

4. A drop of ink is placed in a glass of water. After a time, the ink becomes evenly spread throughout the water, although the water was not stirred. Is this an example of a decrease or an increase in entropy?

Answers are on pages 264–265.

Lesson 8 Radioactivity

How can one element change into another element?

Until less than one hundred years ago, scientists were quite certain that an element could never change into another element. Then, in 1896, the French scientist Henri Becquerel discovered a property of certain elements that is now called **radioactivity.** Scientists soon learned that radioactive elements can and do change into other elements.

What is radioactivity? Radioactivity is the release of energy and matter that results from changes in the nucleus of an atom. Scientists believe that some atoms that have large nuclei may have an imbalance between protons and neutrons. Such atoms are said to be **unstable** because they have some difficulty in holding the nucleus together. Only atoms that have unstable nuclei are radioactive.

The nucleus of an unstable atom undergoes a process called **radioactive decay.** In radioactive decay, the nucleus releases protons, neutrons, or electrons until a more stable arrangement is reached. Of course, once the number of protons in an atom changes, the atom changes into the atom of another element.

All the elements that have an atomic number greater than 83 are radioactive. A common radioactive element is uranium. When an atom of uranium decays, it changes first into another radioactive element of a lower atomic number, called thorium. The thorium continues to decay until it becomes protactinium, which is also radioactive. The decay continues to alter the atom and to lower its atomic number, until it eventually becomes an atom of the nonradioactive element, lead.

Radioactive elements decay at different rates. The amount of time it takes for half of the atoms in a sample of a radioactive element to decay is called the **half-life** of that element. For example, the radioactive element radium has a half-life of 1600 years. This means that out of a 10-gram sample of radium, only 5 grams of radium will remain after 1600 years. The rest of the radium will have decayed into other elements. After 3200 years have gone by, only one-fourth, or 2.5 grams, of the original radium will remain.

The half-lives of different elements vary greatly. For example, the half-life of uranium is over 4 billion years, while the half-life of polonium is only one-millionth of a second.

Because scientists know the half-life of most radioactive elements, they can use their knowledge to determine the age of many substances. For instance, the carbon that is found in living tissue can become radioactive after the tissue has died. When the amount of carbon in the dead tissue is measured against the amount of carbon in living tissue of the same type, scientists can tell how long the radioactive carbon has been decaying. This process, called **carbon dating,** is relatively accurate for objects that are up to 30,000 years old.

In nature, most radioactive elements decay very slowly. But the decay of radioactive elements can be sped up by artificial means. When the rate of decay is sped up, the amount of energy that is released is greatly increased. This is the principle behind **nuclear energy.** The increased energy released by artificially "enriched" radioactive elements is harnessed by machinery to produce heat, which is then used to generate electricity.

Lesson 8 Exercise

Items 1 and 2 refer to the periodic table in Lesson 3.

1. Which of the following elements is most likely to be radioactive?

(1) hydrogen
(2) carbon
(3) nickel
(4) lead
(5) plutonium

2. Uranium emits a particle that consists of two neutrons and two protons when it decays. This is the nucleus of which element?

3. Why do you think that the nuclear fuel in a nuclear reactor must constantly be replaced?

4. In using carbon dating to determine the age of an object, the greatest error would occur when attempting to date which of the following?

(1) a bone that is suspected to be about 14,000 years old
(2) a piece of pure quartz
(3) a very small seashell
(4) the remains of a plant

Answers are on page 265.

Chapter 4 Quiz

1. What would be the net charge of an atom that has ten protons, 12 neutrons, and 11 electrons?

2. A quart of milk normally contains 1,000 units of vitamin A, a molecule necessary to human health. However, if a closed quart of milk is left in a sunlit window, the vitamin A content decreases drastically. Which of the following statements best explains this decrease?
   (1) The vitamin A evaporated.
   (2) The vitamin A dissolved.
   (3) The vitamin A was broken down through a chemical reaction.
   (4) The milk evaporated.
   (5) The vitamin A and milk formed a mixture of uniform consistency.

Items 3 and 4 refer to the following figure.

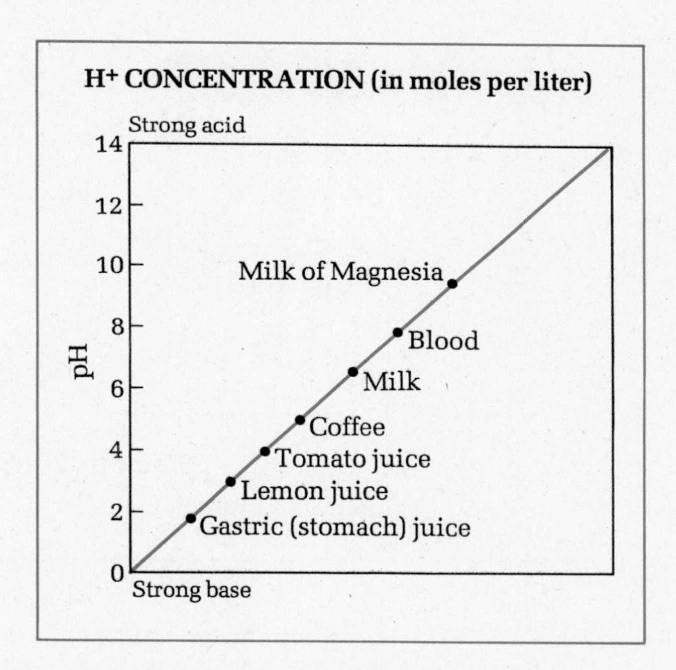

3. Between which two substances would you place water (pH 7.0)?

4. Which of the foods on the graph would probably be most harmful to someone suffering from an "acid stomach" (excess stomach acid)?

Items 5 and 6 refer to the following passage.

A catalyst is a chemical which speeds up the rate at which a chemical reaction takes place. In living organisms, catalysts are usually proteins known as enzymes.

5. Papain is an enzyme obtained from the papaya fruit. It is used by cooks to tenderize meat. Papain acts by breaking up the protein fibers in the meat that cause toughness. Which of the following is evidence that papain is a catalyst? Papain

(1) is a protein
(2) is a naturally occurring plant product
(3) acts to break up protein fibers
(4) breaks up fibers more quickly than they would be by just cooking
(5) works at room temperature

6. The following figure describes a chemical reaction. At which point in time is it most likely a catalyst was added?

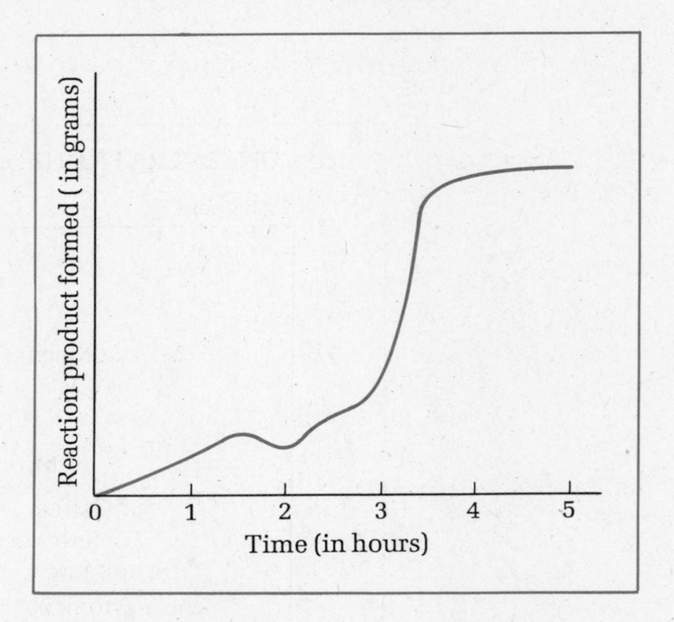

---

7. Acid rain is a phenomenon that occurs when the combustion of fossil fuels such as coal and oil releases large amounts of gaseous sulfur and nitrogen oxides into the atmosphere. These pollutants often travel hundreds or thousands of miles before precipitating back down to Earth, where their high acidity damages both natural ecosystems and artificial structures.

In an effort to combat the effects of acid rain, some governments have begun dumping large quantities of calcium carbonate in lakes to reduce acidity. Calcium carbonate, then, acts as a(n)

(1) acid
(2) base
(3) salt
(4) compound

8. The higher the temperature of a liquid, the more gas can be dissolved in it. A carbonated beverage, such as root beer, is a solution of carbon dioxide gas in water that has flavor added. Why does a soft drink fizz more vigorously when you add ice to it?

9. Consider a hard, solid surface, such as a metal tabletop. Its atoms are made up of nuclei and electrons, but 99 percent of an atom is the empty space between the nuclei and its electron(s). Which of the following is the best explanation for why a table still looks and feels "solid"?

   (1) The electrons of its atoms travel so quickly about their nuclei that they form impenetrable clouds around the entire atom.
   (2) The electrons of its atoms join together to form hard shells around the nuclei.
   (3) The atoms on its outside surface are arranged with their nuclei outward, forming a hard barrier.

10. Of all the types of radiation, gamma radiation is the most penetrating. Alpha particles can be stopped by a thin sheet of paper; beta particles can be stopped by a thin sheet of tinfoil; but gamma rays can penetrate up to 18 inches of lead and up to 32 inches of concrete. Because gamma radiation usually accompanies alpha radiation and beta radiation, extreme precautions must be taken when handling or storing any radioactive substance.

    Based on the information provided, which of the following would probably be the greatest problem in disposing of radioactive wastes from nuclear power plants?

    (1) Radioactive decay rates vary according to the type of material involved.
    (2) In most nuclear power plants, lead and concrete are in short supply.
    (3) Radiation would tend to leak out of any ordinary container in which the wastes would be packaged.
    (4) Nuclear power plants are running out of room to store all the radioactive wastes.
    (5) Radioactive wastes emit more gamma rays than do other types of radioactive substances.

Answers are on page 265.

Chapter

5 Physics

Objective

In this chapter you will read and answer questions about

- Energy, work, and power
- The laws of motion
- The nature of waves
- Electricity and magnetism
- Nuclear energy

Lesson 1 Energy, Work, and Power

What is the relationship between energy, work, and power?

You know that when you are running, you are using energy. You know that when you drive a car, the car uses energy to move. But what exactly is energy? Scientists define **energy** as the capacity to do work. Work is done whenever a force causes an object to move. **Power** is the rate at which the work is done.

Perhaps the most familiar form of energy is **kinetic energy**, or the energy of motion. When a baseball player swings a bat, the bat has kinetic energy. When the bat strikes the ball, the bat does work on the ball by causing it to move. The player who can do this amount of work the fastest is developing the most power.

Objects at rest also have energy, a type known as **potential energy.** Suppose that you have a box of books stored on a closet shelf that is six feet off the ground. If you knock the box off the shelf, it will tumble to the floor and strike anything in its path with considerable force. Before the box fell, it had potential energy; as the box fell, it had kinetic energy.

An important property of energy is that it can change from one form to another. For example, energy is constantly changing from potential to kinetic and back again. However, energy can never be created or destroyed. The idea that energy cannot be created or destroyed is called the law of conservation of energy. According to this law, the total amount of energy in the universe is always the same.

The law of conservation of energy is easier to understand if you consider the following example. When you switch on a light, you can see and feel the light's energy. But when you turn it off, that energy seems to have disappeared. However, it still exists in the form of heat energy in the objects that were warmed by the light when it was on. The energy was not destroyed; it was only converted to a different form.

There are many different forms of potential and kinetic energy. In this chapter, some of the types of energy you will be studying include heat energy, mechanical energy, electrical energy, magnetic energy, light energy, and nuclear energy. Each of these can be converted to another type of energy. Heat energy, for instance, is converted to mechanical energy when steam spins the blades of a giant machine called a turbine. If a turbine is connected to an electric generator, the mechanical energy is converted into electrical energy. Electrical energy can then be converted to light energy in a light bulb. In all these instances, energy is converted without changing the total amount of energy involved.

Lesson 1 Exercise

1. Classify each of the following as exhibiting either kinetic or potential energy.
   (a) a growing plant
   (b) a rotating galaxy
   (c) a dish on a table
2. In a solar cell, sunlight strikes the cell and causes an electric current to flow. This current can then be used to operate a device such as a calculator. Explain how energy is changed from one form to another in this process.
3. Which person is using the most power, a person walking up stairs or a person running up stairs?
4. Which of the following statements is NOT supported by the passage?
   (1) In order to do work, an object must have energy.
   (2) At the point at which an object is dropped, its energy changes from potential energy to kinetic energy.
   (3) As energy is converted from one form to another, some of the energy is always lost.
   (4) Electricity can be generated as energy changes from one form to another.

Answers are on page 265.

Lesson 2 Motion

What are the forces that affect motion?

In order for kinetic energy to do work on an object and move it, some sort of force must be applied to that object. It is this force that actually causes the object to move. Some forces that commonly act on us are gravity, friction, and inertia. **Gravity** is the force every object in the universe exerts on every other object. **Friction** is the force that opposes motion, or the force of resistance. **Inertia** is the tendency of matter to remain at its present state of motion.

In the example of the car in the previous lesson, it is the force applied by the wheels against the ground that channels the engine's mechanical energy. The strength and direction of that force determines how the car moves. To get the car moving, inertia would have to be overcome. If the car were then driven up hill, gravity would be pulling against the direction of motion. At the same time, friction between the wheels of the car and the road would also be slowing the car's motion. As the car moved, inertia would tend to keep the car moving, in this case working with the force being applied. To stop the car, however, the driver would have to apply the brakes. The force of inertia would have to be overcome to stop the car.

In an attempt to describe and predict what sort of energy and force causes objects to move and to stop moving, Isaac Newton, a seventeenth century scientist, set down three laws of motion.

Newton's first law of motion is also known as the law of inertia. The law states that an object at rest will tend to stay at rest, and that an object in motion will tend to continue in motion along a straight line, unless acted upon by an outside force.

If you take a ride on a roller coaster, you can feel inertia at work. As the car begins to move, you are pressed backward into your seat. This happens because your inertia works to keep you at rest. If the car stops abruptly, you will continue moving forward. Once you're moving, your inertia works to keep you in motion. Inertia also explains why a satellite launched into orbit would fly off into space if the force of gravity did not keep it moving in a circular path.

Newton's second law states that the force needed to move an object is directly proportional to an object's mass. This means that the greater the mass of an object, the more force must be applied to change its state of motion. If you have ever tried to move a piano, you know that it takes considerably more energy to move it than to move a guitar. As you might expect, it also takes more force to stop a larger object than to stop a smaller one.

Newton's third law of motion states that for every force, there is an equal and opposite force. You can see how this principle works if you jump off a diving board or push yourself off from the railing that surrounds a skating rink. The downward push on the diving board results in an equivalent upward push that lifts you into the air. The push against the railing sends you out into the skating rink in the opposite direction of the push. Rockets take advantage of this principle. The burning of fuel pushes gases out the rear of the rocket, and the rocket is thrust forward into space.

Lesson 2 Exercise

1. Which force are you applying when you apply brakes to a car's wheels?
2. Which of Newton's laws explains why seat belts prevent injury in automobiles? Give a reason for your answer.

3. Explain in terms of Newton's second law why a small car uses less fuel than a larger car.

Answers are on pages 265–266.

Lesson 3 Heat and Mechanical Energy

How are heat and mechanical energy related?

One important form of kinetic energy is heat. Since kinetic energy is the energy of motion, it may seem strange that heat is also kinetic energy. After all, a hot object is not necessarily in motion. However, heat is a form of kinetic energy on an internal, or molecular, level.

Molecules in an object are in constant motion. The molecules in a solid move more slowly than do the molecules in a liquid or a gas, but they are still moving. When heat energy is applied to molecules, they gain energy and move faster. The faster the molecules move, the hotter they seem to be. Therefore, an object's temperature actually refers to the kinetic energy of its molecules.

Like all forms of energy, heat has the ability to do work. In order to do work, heat must be converted to mechanical energy.

A familiar example of the conversion of heat to mechanical energy is a running car engine. As the engine burns fuel, the hot gases it produces do work on the pistons, which move up and down. This mechanical energy is transferred to the wheels of the car, which propel it forward. The heat energy that is not converted to mechanical energy is expelled by the engine in the hot exhaust fumes. Mechanical energy can also be converted to heat energy through friction. On a cold day, you can rub your hands together to keep them warm. The work of moving one hand against the direction of the other produces heat. Friction also produces heat in many types of machines. For example, the print head of a computer printer becomes hot as the printer moves rapidly across the page. The blades of an electric blender will feel warm after they have been in use for several minutes.

As you can see, heat energy and mechanical energy are strongly related. The study of the connection between these two forms of energy is called **thermodynamics.**

The study of thermodynamics is governed by two important scientific laws. The first law of thermodynamics is the law of conservation of energy. In the example of the car engine above, you saw this law in action as the unconverted heat energy was not lost, but was expelled into the air.

The second law of thermodynamics states that heat will always move from a hotter object to a cooler one. This law becomes obvious if you touch a hot stove—heat from the stove will quickly move to your hand. Note that the opposite is not true. Heat will not flow on its own from a cold object to a hot object.

Lesson 3 Exercise

1. How are heat and work related?
2. A person slides too quickly down a rope and experiences a rope "burn." Explain why this happens.
3. An ice cube is placed in a cup of hot tea. In which direction will heat flow between the ice cube and the tea?
4. Which of the following statements best explains why oiling the moving parts of an automobile increases the amount of mechanical energy available to move the car?
   (1) The oil lowers the temperature of the hot gases produced by burning fuel.
   (2) The oil decreases the friction between the moving parts.
   (3) The oil allows the engine to operate at cooler temperatures.
   (4) The oil decreases the friction between car tires and the road.

Answers are on page 266.

Lesson 4 The Nature of Waves

What properties do all waves have in common?

One of the most familiar kinds of motion is the motion of waves. You can see ocean waves as they move across the surface of the water. But there are many other types of waves around us that are not so obvious, such as radio waves and sound waves, because they are invisible.

A wave itself is not a form of energy. A wave is a way for energy to be transferred from one place to another. You can understand how energy is transferred by watching a speedboat moving through the water. Kinetic energy from the speedboat creates waves which carry that energy away from the boat. If there is a smaller boat nearby, you can see the effect of the transferred kinetic energy. The smaller boat will begin to rock as waves transfer the speedboat's energy. In this case, the transferred energy was kinetic energy. Similarly, a sound wave carries sound energy, a light wave carries light energy, and so on. Energy that is carried on a wave is called radiant energy.

All waves have a characteristic shape and the following properties: amplitude, wavelength, and frequency. It is the combination of these properties in different proportions that determines the form of a wave.

Amplitude is the maximum distance a wave vibrates from its normal rest position. The amplitude of an ocean wave would be its highest point

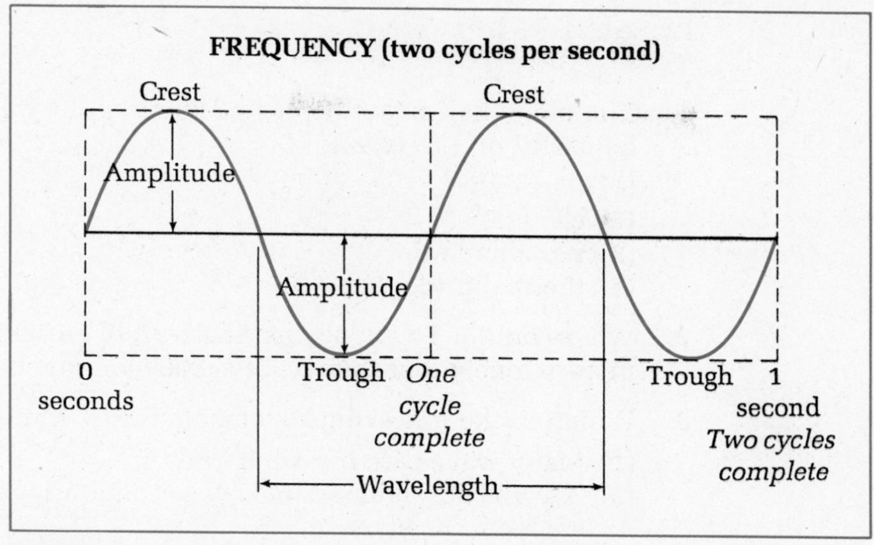

FIGURE 1

above or its lowest point below the calm surface of the ocean. The high point in a wave is called the crest. The low point in a wave is called the trough. The distance from rest position to the crest will always equal the distance from rest position to the trough.

The distance between one crest and the next is called the **wavelength.** The number of waves that pass a given point in a certain amount of time is called the **frequency** of a wave. For example, if two ocean waves, measured from crest to crest, pass a buoy in one second, the frequency of the wave is two per second.

A wave of very long wavelength has a low frequency, because it will take that wave longer to pass a given point. Such low-frequency waves are low in energy. Familiar examples of low-energy waves are light waves and microwaves. Likewise, if a wave has a very short wavelength, it takes less time to pass a given point, and it therefore has a high frequency. High-frequency waves such as X-rays and gamma rays are high-energy waves. A gamma ray, produced during nuclear radiation, has so much energy that it can pass through almost any material.

The substance through which a wave travels is called the **medium.** When a wave travels through the atmosphere, air is the medium. Certain types of waves, such as sound waves, must have a medium through which to travel—that is why sound cannot travel in a vacuum. Other types of waves, such as light waves or radio waves, do not need a medium through which to travel. Thus light can reach Earth from outer space, where there is no atmosphere.

Waves travel at different speeds through different mediums. Sound waves travel faster through solids than through liquids or gases. Perhaps you have heard the old saying "Put your ear to the ground." The pioneers knew that the sound of approaching horses would travel much faster through the ground than through the air.

Lesson 4 Exercise

1. The moon has no atmosphere. Which of the following items would NOT be useful on the moon?
   (1) flashlight
   (2) camera
   (3) portable radio
   (4) thermal underwear

2. Why would a TV set against the wall of an apartment be more likely to disturb a neighbor than a TV set several feet from the wall?

3. Which of the following statements is NOT implied in the lesson?
   (1) Many waves are invisible to us.
   (2) A wave may transfer energy without any movement of matter from source to receiver.
   (3) A single wave can travel hundreds of miles.
   (4) Sound waves travel faster than light waves.
   (5) The longer the wavelength of a wave is, the lower it is in energy.

4. In a light wave, a greater amplitude generates a brighter light. Assuming a change in amplitude causes a similar change in sound, would a greater amplitude generate a louder sound or a quieter sound?

Answers are on page 266.

Lesson 5 Behavior of Light

What are the properties of light?

Electromagnetic radiation is made up of waves that can travel through a vacuum at the speed of light—186,000 miles per second. Different types of electromagnetic radiation carry different degrees of radiant energy, and have different frequencies. Scientists arrange all the frequencies in a scale from the highest to the lowest. This scale is called the **electromagnetic spectrum.**

Most of the waves in the electromagnetic spectrum are invisible to our eyes. These include X-rays, radiowaves, and microwaves. But there are certain waves that we can see. We know these electromagnetic waves as visible light.

Like sound waves and water waves, light has the property of **reflection.** When light strikes an object such as a red rose, chemicals in the object absorb most of the colors in the light. But red light is reflected back to you, so the rose appears red. You can see yourself in a mirror because of reflection. In fact, we wouldn't be able to see at all were it not for this property of light, because light reflects off nearly every object it touches. This is why in a very dark room you not only can't see colors of the objects around you, you can't see the objects themselves.

Light waves also have a property called **refraction.** When a beam of light passes from one medium to another, the light is bent, or refracted. Refraction occurs because light, like all waves, travels at different speeds through different mediums. The amount that the light is refracted depends upon the particular materials involved. It is because of refraction that an object underwater often looks distorted when viewed from the water's surface. Starlight that is refracted when passing through Earth's atmosphere makes the stars appear to twinkle.

We know the different frequencies of visible light as the colors of the rainbow: red, orange, yellow, green, blue, and violet. In sunlight these colors are mixed together and we see them as white light. Each frequency refracts, however, at slightly different angles. This means that when it passes through certain mediums, such as glass or water, each frequency is separated from the others. This is what happens when light passes through a prism. A prism breaks up the frequencies of white light into the colors of the visible spectrum. Similarly, a rainbow is formed in the sky by the refraction of the sun's rays in falling rain or in mist.

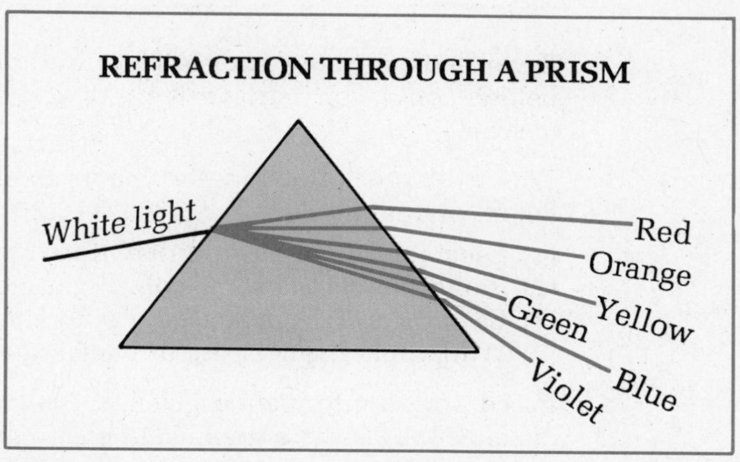

FIGURE 2

Lesson 5 Exercise

Item 1 refers to the following figure.

1. Which property of light does the figure illustrate—reflection or refraction?

---

2. If you looked into a pool of clear water and saw a gold ring directly below, could you retrieve it by diving straight down? Explain your answer.

3. Which of the following statements explains why you see different colors in a rainbow?

(1) Each color light has a different frequency.
(2) Raindrops refract some color light more easily than others.
(3) Each color light reflects at a different speed.
(4) White light cannot pass through water droplets.

4. Based on the information given in the lesson, which of the following conclusions can be drawn about electromagnetic radiation?

(1) The sun does not emit electromagnetic radiation.
(2) Electromagnetic radiation may be either reflected or refracted.
(3) Electromagnetic radiation is not absorbed by the objects on which it falls.
(4) Violet light has more energy than any other type of electromagnetic radiation.
(5) Most types of electromagnetic radiation cannot pass through water.

Answers are on page 266.

Lesson 6 Electricity

Why is electricity a basic property of all matter?

Electricity is an important form of energy in our daily lives. We usually think of electrical energy as the force that runs our machinery and makes our light bulbs work. That is one form of electricity. But electricity is actually a basic property of all matter.

Recall that all matter is made up of atoms, and that all atoms contain protons that have positive charges and electrons that have negative charges. Because opposite charges attract each other, it is the positive charges and the negative charges within an atom that hold it together. Protons, however, are fixed within the atom, while electrons are free to move. It is the movement of electrons between atoms that causes electricity.

The most common type of electricity found in nature is called **static electricity,** which results from the buildup of charges on objects. You can see static electricity at work if you dry your clothes in a dryer. As the clothes rub against each other, they exchange electrons. Some build up positive charges by losing electrons, while others build up negative charges by gaining electrons. The opposite charges cause the clothes to cling to each other.

If the exchange of electrons continues, the charge will become strong enough to pull the electrons back to the positively charged object. The electrons jump from one object to another. This type of discharge is called a spark. Lightning is actually a giant spark that results from the sudden transfer of electrons between two clouds or between a cloud and Earth.

The movement of electrons that produces static electricity is uncontrolled. In order for electricity to power appliances and to do other useful work, it must be controlled. **Current electricity** is the controlled flow of electrons through an **electric current.**

An electric current is created by providing electrons with **electric potential energy.** To give an electron potential energy, work must be done to move it away from its atom. The electron now possesses potential energy. The electron then moves back toward the positive nucleus from which it has been separated, thus releasing its potential energy. The electric current itself moves from a point with negative charges toward a point with positive charges.

To make use of an electric current, a device must be able to separate an electron from its atom, and to then make use of the resulting energy. Most devices use electricity to produce heat, light, or motion. In a heating element or light bulb, electrons collide with the atoms that make up the heating element or lighting filament. Other electrons continue flowing from behind them, pushing the first electrons forward. A sort of "electrical friction" is created as all the electrons make their way through the appliance. As they move, the electrons transfer much of their energy to the atoms through which they are pushing. The energy increase causes the atoms to move back and forth rapidly. As the atoms move, they give off heat and light.

Lesson 6 Exercise

1. What is the main difference between static electricity and current electricity?
2. If you scuff your feet across a wool carpet and then touch a metal doorknob, you will probably get a shock. Explain why.
3. About 90 percent of the electrical energy that is used by an electric coffee pot is converted to heat energy in heating the water in the pot. Which of the following statements best explains what happens to the other 10 percent of the energy?
   (1) The other energy is destroyed.
   (2) The other energy is released to the environment in a different form, such as light.
   (3) The other energy is still in the form of potential energy.

Answers are on page 266.

Lesson 7 Magnetism

How is magnetism related to electricity?

Magnetism, like electricity, is a property of all matter. Scientists believe that magnetism is caused by the rotation of electrons within a substance. Every electron in every substance acts as a magnet. Because all matter is made of atoms that contain electrons, all matter is capable of exerting magnetic force.

In most substances, the electrons rotate in many different directions. For this reason, the magnetic force of each atom is canceled out by the others. But in certain substances, the electrons all rotate in the same direction. Their magnetic forces line up to create a magnetic field. These substances are called magnetic, and can exert strong magnetic force on other objects.

Magnetism is a force of pull, or attraction, and push, or repulsion. Every magnet has two ends, called **poles,** from which the magnetic force extends. One pole is the magnet's north pole, and the other is its south pole. The magnetic force, in the form of a **magnetic field**, extends outward from the poles to create an area around the magnet where its magnetism can be felt.

If two magnets are brought near each other, they will exert a force on each other. If the north pole of one magnet is brought near the south pole of another magnet, the two poles will attract each other. If the north poles of two magnets are brought near each other, they will repel each other. The same thing will happen if two south poles are brought together. These forces of attraction and repulsion are summed up by this rule: Opposite poles attract each other, while like poles repel each other.

Earth itself acts as a giant magnet, with its north and south poles near the geographical poles. Because Earth's magnetic field is so strong, it exerts force on every magnet on the planet. This is how compasses work. The magnetized needle on a compass is attracted by Earth's poles and lines up with its magnetic field.

An important relationship exists between magnetism and electricity. This relationship is called **electromagnetism.** Electromagnetism involves the generating of electricity from a magnetic field or vice versa. When a wire is passed across a magnetic field, electrons in the wire will move, and an electric current flows through the wire. In the electric generators found at power plants, electricity is created by moving huge magnets past coils of copper. The electricity generated in copper is then carried to surrounding areas by power lines.

Lesson 7 Exercise

Item 1 is based on the following diagram.

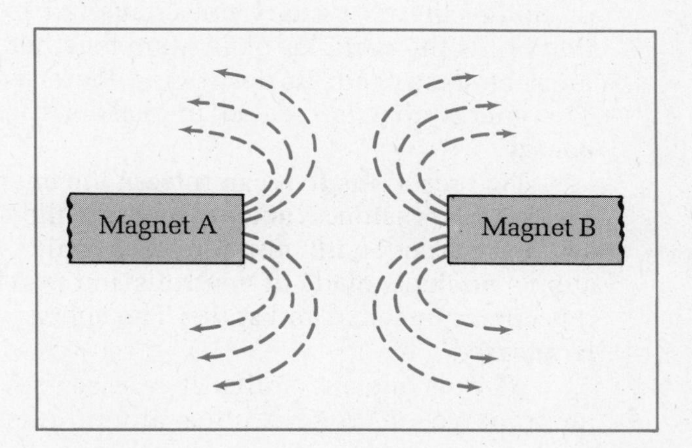

1. The above diagram shows the magnetic force lines that exist between magnet A and magnet B. Which of the following relationships could exist between magnet A and magnet B?

(1) The magnets are far apart from each other.
(2) The north pole of magnet A is close to the south pole of magnet B.
(3) The south pole of magnet A is close to the north pole of magnet B.
(4) The north pole of magnet A is close to the north pole of magnet B.
(5) Magnet A is much stronger than magnet B.

2. The point on a compass marked "north" points toward Earth's north pole. Taking into account the rule of magnetic attraction and repulsion, form a hypothesis as to why.

3. A material that allows current to flow freely is called a conductor. Such a material has atoms with electrons that are free to move from atom to atom. Materials that do not have free-moving electrons and thus do not let current pass are called insulators. Based on the use of copper in electric generators, is copper a conductor or an insulator?

Answers are on pages 266–267.

Lesson 8 Nuclear Energy: The Energy of the Atom

How is matter converted into energy?

We have seen how energy can be converted to other forms, and how the total amount of energy always remains the same. However, there is one form of energy that is usually not converted to other forms. This is the energy that holds the particles of an atom together. To convert this energy, the atom must be destroyed. In destroying the atom, matter is converted to energy. The energy that is created in the destruction of matter is called **nuclear energy.**

Two reactions that can release the energy in an atom are nuclear fission and nuclear fusion. **Nuclear fission** is the splitting apart of an atom. Atoms are bombarded with minute, subatomic particles called **neutrons.** As an atom's **nucleus,** made of neutrons and protons, absorbs the excess neutrons, it becomes unstable and splits. The energy which holds the nucleus together is released.

As one nucleus splits, it releases several additional neutrons. These neutrons go on to split other atomic nuclei, which in turn release more neutrons, and so on. The result is what is called a **nuclear chain reaction.** If the chain reaction is not controlled, an incredible amount of energy is released in a very short time. This is what happens in a nuclear explosion.

Nuclear reactions are controlled in devices called **nuclear reactors.** Nuclear reactors are the central sources of energy at nuclear power plants. In a nuclear reactor, the speed of a chain reaction is controlled by substances that absorb excess neutrons. Reactors also have a cooling system to prevent overheating and to carry away the heat energy that is generated by the reaction. This heat energy is then used to generate electricity.

Another type of nuclear reaction that releases a great deal of energy is nuclear fusion. **Nuclear fusion** is the joining of two atomic nuclei to produce a larger atomic nucleus. On the surface of the sun and other stars, hydrogen atoms fuse to produce helium atoms and energy. Thus it is nuclear fusion that produces the sun's radiant energy.

Scientists have been trying for many years to produce controlled fusion reactions on Earth. So far they have been successful only in producing uncontrolled fusion reactions, using hydrogen bombs. If fusion can ever be controlled, it might prove to be a very useful source of energy. Because it does not produce as much radioactive material as fission, it would also be much safer.

Lesson 8 Exercise

1. In what ways are fission reactions and fusion reactions similar? In what ways are they different?
2. Which of the following situations is most like a nuclear chain reaction?
   (1) One person telephones two other persons, who each in turn telephone two other persons, who each in turn telephone two other persons.
   (2) A car stops suddenly, causing all the cars behind it to crash.
   (3) One domino falls over and knocks down all the dominoes behind it.
   (4) A person eats half a bowl of peanuts in 20 minutes, then eats half of what is left in the next 20 minutes.
3. It can be inferred from the passage that radiation from nuclear power could escape to the environment in what ways?

Answers are on page 267.

Chapter 5 Quiz

Item 1 refers to the following figure.

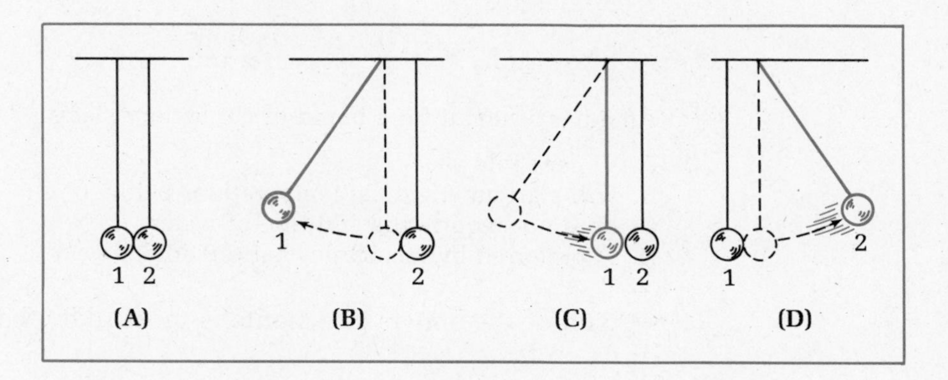

1. Which of Newton's laws of motion is being violated in the figure? Explain why.

2. You are driving down a road just after a heavy rainfall. The sun has come out and the water on the road forms a shimmering surface. You cannot see where you are going because of the glare. Which of the following is the best explanation for this situation?

   (1) The water on the road acts like a mirror that reflects most of the sunlight back into your eyes.
   (2) The water on the road is heated by the sunlight, producing heat waves that reduce visibility.
   (3) Sunlight is refracted by water droplets into your eyes.
   (4) The black tar surface of the road absorbs most of the sunlight, reducing visibility.
   (5) The car's tires spray the water up as a mist that reduces visibility.

3. Which of the following activities represent doing work to a physicist?

   (a) solving a math problem
   (b) hitting a golf ball
   (c) pushing a shopping cart
   (d) standing in line at the bank
   (e) rolling a ball
   (f) washing dishes

4. Which one of the following statements is evidence that electricity and magnetism are interrelated?

   (1) Electromagnetic radiation does not require a medium through which to travel.
   (2) The magnetized needle on a compass lines up with Earth's magnetic field.
   (3) The electric current flowing through electric power lines creates a magnetic field.
   (4) Electrons tend to move from an area having a negative charge to that having a positive charge.
   (5) An electric appliance can produce heat and/or light.

5. When you bake cookies, energy in the form of heat is absorbed by the cookies. Baking is an application of which law of thermodynamics?

Items 6 and 7 refer to the following information.

An advertisement for a brand of car battery states

A. dependable year-round
B. will start an engine at temperatures below 0° F
C. does not require the addition of water
D. is preferred by mechanics nationwide

6. Which of the above statements is most likely based on opinion rather than on facts?

   (1) A only
   (2) B only
   (3) A and C only
   (4) B and C only
   (5) A and D only

7. Which of the above statements can be tested experimentally?
   (1) A only
   (2) B only
   (3) A and C only
   (4) B and C only
   (5) A and D only

Items 8 and 9 are based on the following diagram.

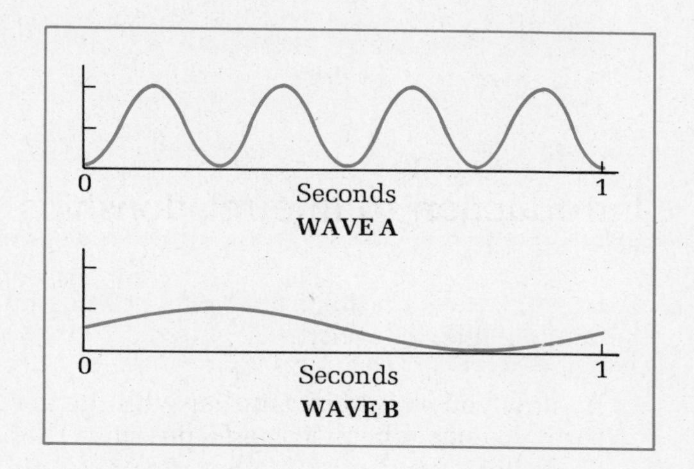

8. Which of the above waves has the greater amplitude? the greater wavelength?

9. Which of the above waves would you expect to have more energy? Why?

---

10. Give evidence to support or refute the following statement: Unlike nuclear fission reactions, nuclear fusion reactions occur in nature.

Answers are on page 267.

Chapter 6

Interrelationships Among the Sciences

Objective

In this chapter you will read and answer questions about

- interrelationships among the sciences
- applications of the sciences to real world problems

Lesson 1 Introduction to Interrelationships

In what ways do the branches of science interrelate with each other and outside fields of study?

By now you should be familiar with the major branches of science: biology, earth science, chemistry, and physics. These branches are all separate but dependent upon one another. For example, the inner workings of a cell involve chemical reactions such as cellular respiration or synthesis. A biologist could not truly understand a cell without a working knowledge of the chemistry of living things, or **biochemistry.** Although some scientists specialize in certain fields that may be considered apart from other areas, many times scientists from different branches work together to achieve a common goal. In the space program, chemists, physicists, and biologists have all contributed to the study of other parts of the universe, a subdivision of earth science.

The lessons that follow illustrate a few of the connections between the sciences. As you read them, keep in mind that science is a method for uncovering truths about our world. All the branches of science have the scientific method in common. Keep in mind also that the scientific method may be applied to fields other than science. Athletic performances, for example, have improved through the application of physics principles to sports. Finally, consider that science is also influenced by other fields. Scientists use the skills of mathematicians to analyze data. Scientists also depend on people who can help them communicate their findings to the general public. In these and many other ways, the branches of science interconnect not only with each other, but with the rest of the world.

Lesson 1 Exercise

Items 1 to 4 refer to the following passage.

The fossil of a fish is found on a mountaintop 10 miles from the nearest body of water. How can the presence of the fish fossil be explained? According to one scientist, 20,000 years ago the mountain was under water. A second scientist is called in to determine if the fish is old enough to have been there at the time the mountain was still under water. Through carbon-14 dating, which makes use of the knowledge of radioactive carbon's half-life to determine the period of time to which a fossil belongs, the scientist estimates the fossil is approximately 25,000 years old. A third scientist is studying the ecology of the area from 100,000 years ago to the present. This scientist identifies the species of fish to which the fossil belongs, and confirms that this species was abundant in the area thousands of years ago.

1. Based on the information in the passage, construct a hypothesis explaining the presence of the fish fossil on the mountaintop.

2. Which scientist proved that the fish was not brought to the mountain by a feeding bird 5000 years ago?

3. Which branches of science are represented by the three scientists who provided information about the fish fossil?

4. The first scientist believes that the mountain is growing at the rate of six inches per year. How does the fish fossil support this theory?

Answers are on page 267.

Lesson 2 Space Exploration

How are chemistry, physics, astronomy, and biology involved in space exploration?

Space exploration draws on knowledge of physics, chemistry, biology, and astronomy, as well as requiring specialized research of its own. Rocket engineers must design rockets that can escape Earth's gravitational force. A rocket illustrates Newton's third law of motion. The rocket burns fuel during a chemical reaction. The resulting gases blast out of the rocket's rear at very high speeds, creating the force necessary to push the rocket away from Earth. Once in outer space, rockets are supplied with cameras, computers, and other equipment to gather information about the solar system and other areas of the universe. This information is of great value to earth scientists.

Knowledge of life processes is useful in manned space exploration. People need oxygen, food, and water for life. They also require a safe way to dispose of wastes. Biologists, working with other scientists and engineers, help to solve these problems.

For shorter trips, machines recycle air and water, removing the waste products and returning them for reuse. Food is brought along from Earth. For longer trips planned in the future, a more permanent solution is necessary. On Earth, plants supply oxygen and water and provide food. Biologists are studying the use of plants in space to maintain a liveable environment over a long period of time. In an experiment conducted in the Soviet Union, two men lived for five months inside a thirty-by-forty-foot room. All purification of air and water was done by plants. An additional garden supplied some of the men's food.

If plants are to be grown in a space station, biologists with knowledge of human nutrition would need to advise the gardeners. A "space garden" would probably include sources of protein, such as beans; sources of fat, such as nuts; and sources of carbohydrates, such as potatoes and wheat. It would also need to provide the vitamins and minerals necessary for human health. Otherwise, astronauts on long trips might develop one or more of the diseases that come from poor nutrition.

Lesson 2 Exercise

1. Which branch of science would study the reactions of people to the physical stresses encountered in space, such as weightlessness and radiation exposure?

(1) chemistry
(2) physics
(3) biology
(4) astronomy

2. Does the lesson provide evidence that plants could be used on long space voyages to maintain a liveable environment for human beings? Explain your answer.

3. In planning a garden for a space station or long space voyage, space scientists have considerations in addition to nutrition. For example, they have to consider how fast a particular plant grows: A tree that requires five years to bear fruit would be useless as a food supply on a one-year voyage. Keeping in mind the constraints of a spaceship environment, name at least one other consideration in choosing plants for a garden in space.

Answers are on pages 268–268.

Lesson 3 Applying Science to the Problem of Air Pollution

How can scientists help control air pollution?

Air pollution is the result of both natural phenomena and human activities. A volcano and a factory smokestack, each pouring poisonous gases and particles of soot and ash into the air, are both sources of pollution.

Air pollution affects both human health and the environment. Long-term exposure to air pollution has been linked with lung cancer, bronchitis, asthma, and heart disease. Air pollution affects the environment in several ways. Smoke containing chemical wastes can be absorbed into the atmosphere and produce acid rain, which lowers the pH of lakes and streams. When the water in which they live becomes too acidic, fish and other living things die. Acid rain also damages trees, by stripping them of the waxy coating that protects against disease and water loss. Some farm crops are stunted by acid rain.

Particles in the air, another form of pollution, may be raised to a dangerous level by storms or other atmospheric disturbances, or by volcanic eruptions. The particles absorb and scatter sunlight, and they reduce the amount of light reaching Earth. This causes a drop in temperature, which can affect climate.

Reducing air pollution and its effects depends on the work of different kinds of scientists. Earth scientists who specialize in predicting the weather can warn of times when pollution is going to be bad. For instance, an **atmospheric inversion** can cause a buildup of pollutants. An atmospheric inversion is a condition in which an upper layer of warm air acts as a "lid" to prevent the movement upward of cooler air at the surface. This condition occurs often in the Los Angeles Basin, which is sheltered on two sides by mountains. If communities are warned of an atmospheric inversion in advance, they can lessen pollution by reducing the use of polluting fuels for the period of the inversion.

Biologists also work at finding solutions to the effects of air pollution. For example, they have attempted to make lakes temporarily less acidic by adding large amounts of lime to the lakes. The lime acts as a base to neutralize the water. Studies by biologists have also been important in educating people about the dangers of air pollution.

Engineers apply science to design filters to reduce air pollution from cars, power plants, and factories. They have invented scrubbing systems to clean high-sulfur coal, a major cause of acid rain. They also have created devices to detect air pollution.

Lesson 3 Exercise

Items 1 and 2 refer to the following table.

Sources of Particulate Pollution in the United States

Source	Emissions (millions of tons per year)
Natural dusts	63
Wildfire	37
Controlled slash burning	6
Controlled burning of forest litter	11
Agricultural burning	2.4
Transportation	1.2
Incineration	0.931
Rubber from tires	0.300
Cigarette smoke	0.230
Aerosols from spray cans	0.390
Ocean salt spray	0.340

1. Which of the following is the largest source of particulate pollution?

(1) natural dusts
(2) wildfire
(3) transportation
(4) incineration
(5) cigarette smoke

2. Which would reduce air pollution more—switching to bicycle transportation or stopping the burning of forest litter?

---

3. In London in 1952, 4000 people died from a sudden buildup of pollutants. Suggest an explanation for this sudden buildup.

4. In 1816, a major volcanic explosion by Mount Tambora in Indonesia caused it to snow in New England in June and resulted in frosts through August. Why did this explosion cause cold temperatures in New England?

Answers are on page 268.

Chapter 6 Quiz

1. In studying how blood flows in and out of the heart, biologists must consider the speed and pressure of blood at different points in the body. The biologists need a knowledge of which branch of science to perform their calculations?
   (1) biology
   (2) chemistry
   (3) earth science
   (4) physics

2. At the start of a race, the runners push hard against their starting blocks and against the ground to which the blocks are attached. At the same time, the blocks and the ground push back against the runners. Earth is so large compared to the runners that it hardly moves when they push. Thus nearly all of the force is concentrated in the runners, sending them off down the track. Why would a physicist be interested in this description of the start of a race?

Items 3 to 7 refer to the following passage.

A typical meal of chicken, potatoes, salad with dressing, and milk contains the basic nutrients. There are proteins in milk and chicken, carbohydrates in the potatoes, salad, and milk; and fats in the chicken, salad dressing, and milk. There are vitamins and minerals in all the foods. But to liberate the nutrients in foods so that the body's cells can use them, the foods must be digested.

The digestive system breaks down food into a usable form. Digestion begins in the mouth. As you chew your dinner, saliva moistens it and begins to break the starchy complex carbohydrates apart into simple sugars. From the mouth the food passes through the esophagus to the stomach. The stomach further breaks down food by kneading and churning it. The acids in the digestive juices begin to break apart proteins into their amino acid components.

Within a few hours, most of the food passes into the small intestine, the body's major digestive organ. Here, food is reduced to simple molecules that can pass through the wall of the small intestine and on into the bloodstream. The pancreas helps by pouring digestive liquids into the small intestine. The liver secretes bile, a liquid that is first stored in the gallbladder and then released into the small intestine. Bile breaks up fats.

As digestion is completed, blood from the digestive tract pours into the liver. The liver cells transform fatty acids into usable forms. They make amino acids into blood proteins and other products needed by the body. They rearrange sugar molecules into glucose, or blood sugar. They also remove poisons, such as alcohol. From the liver, nutrients flow into blood to the body's cells. Any unusable substances pass from the small intestine to the large intestine and on out of the body.

3. Define the word *digestion*.

4. Mechanical digestion requires movement of parts of the body. Chemical digestion requires the breaking apart and dissolving of food molecules. Classify each of the following as either mechanical digestion or chemical digestion.

(a) chewing food
(b) digesting proteins into amino acids
(c) kneading and churning food in the stomach
(d) the action of bile on fats

5. If you hold a piece of soda cracker in your mouth for a time, it will begin to taste sweet. Is the sweet taste evidence that a soda cracker contains complex carbohydrates? Explain your answer.

6. Which of the following statements about digestion would most interest a biologist?

(1) Food is broken apart with the aid of digestive juices and special chemicals called enzymes.
(2) The final products of digestion are nutrients in a form the body's cells can use.
(3) Muscles in the digestive tract contract in rhythmic waves to move food from one part of the system to the next.
(4) Saliva is a liquid composed of water and enzymes.
(5) Villi are small fingerlike projections in the lining of the small intestine that increase its surface area.

7. Metabolism is the process by which the body's cells change nutrients into substances they need. Which two branches of science would be most interested in metabolism?

Items 8 to 10 refer to the following figure.

THE WIND-CHILL FACTOR

Temperature (Fahrenheit)

Wind (m.p.h.)	35	30	25	20	15	10	5	0	−5	−10	−15
Calm	35	30	25	20	15	10	5	0	−5	−10	−15
5	33	27	21	16	12	7	1	−6	−11	−15	−20
10	21	16	9	2	−2	−9	−15	−22	−27	−31	−38
15	16	11	1	−6	−11	−18	−25	−33	−40	−45	−51
20	12	−3	−4	−9	−17	−24	−32	−40	−46	−52	−60
25	7	0	−7	−15	−22	−29	−37	−45	−52	−58	−67
30	5	−2	−11	−18	−26	−33	−41	−49	−56	−63	−70
35	3	−4	13	−20	−27	−35	−43	−52	−60	−67	−72
40	1	−4	−15	−22	−29	−36	−45	−54	−62	−69	−76

8. The National Weather Service issues a special advisory when the wind-chill factor falls below −35° F. Which of the following situations would cause a special advisory to be issued?

(1) temperatures of 30° F with winds from the north
(2) temperatures of 30° F with 5 mile-per-hour winds from the east
(3) temperatures of 5° F with 15 mile-per-hour winds from the west
(4) temperatures of 10° F with 40 mile-per-hour winds from the south
(5) temperatures of 5° F on a calm day

9. Which one of the following phenomena does the wind-chill factor explain?

(1) why the sunny side of a street feels warmer than the shady side
(2) why antifreeze keeps a car's engine from freezing on a cold day
(3) why it feels colder on a day with strong winds than it does on a calm day of the same temperature
(4) why extreme cold and strong winds are dangerous to living things
(5) why many plants in North America enter a period of dormancy during the winter months

10. The wind-chill factor was calculated by two scientists in Antarctica, who conducted experiments measuring the rate at which water freezes in plastic containers. What error might there be in applying their results to human flesh? Explain your answer.

Answers are on page 268.

Unit II Test

Items 1 and 2 are based on the following information.

Changes in which the physical properties of a substance are altered, but the substance remains the same kind of matter, are called physical changes. Changes in which the substances turn into other substances with different properties are called chemical changes.

1. Solid carbon dioxide (dry ice) is placed in water and bubbles up as carbon dioxide gas. This is a ____________ change.

2. Which of the following is NOT an example of a physical change?

(1) melting ice
(2) burning wood
(3) adding sugar to water
(4) brewing tea
(5) shaking up oil and vinegar

Items 3 to 7 are based on the following passage.

Glaciers occur in regions where it is so cold that snow collects year after year without melting during the summer months. Over time, the accumulation of snow reaches depths as great as three miles. As more snow collects on top, the snow on the bottom is compacted into ice by the weight of the additional snow. Eventually, the mass of ice will begin to slide down the mountain where it formed.

As they move, glaciers carry embedded debris, such as huge boulders, rocks, and pebbles. In addition, they push before them and to the side enormous mounds of earth and rubble, called **moraines.** When the glacier retreats, or melts, the moraines are left behind.

There are two kinds of glaciers. **Valley glaciers** form among mountains and creep slowly downward. The rocks and pebbles embedded in the bottom and the sides of the glacier become scraping and gouging tools. They cut long, deep grooves in the underlying rock, parallel to the direction of movement. They scoop out the sides of the valley, so that glaciated valleys have a typical "U" shape. Near the mountain peaks bowl-shaped depressions, called **cirques**, are scooped out. Cirques turn into lakes when the ice melts.

Continental glaciers are glaciers that cover large areas of land. The only such ice sheets on Earth today are Antarctica at the South Pole and Greenland near the North Pole. During the last **ice age**, which ended 10,000 to 12,000 years ago, an ice sheet covered all the territory that is now Canada and part of the United States. Ice also extended much further north from the South Pole than it does now. In the past million years, these ice sheets have advanced and retreated four times. For the hundred million years before that, there appear to have been no continental glaciers.

3. Explain how a glacier forms hills and scoops out valleys as it advances.

4. Which of the valleys in the figure below was probably formed by an advancing glacier?

A

B

5. A New England farmer builds a stone wall out of rocks found in his fields. A local geologist finds that the mineral content of the rocks is different from that of any nearby rock formation. Based on information in the passage, suggest an explanation for this finding.

6. Which of the following is an example of the direct involvement of Earth's gravity in glacier formation or activity?
A. Valley glaciers creep down from between mountain peaks.
B. Ice sheets push mounds of earth and rock before them as they advance.
C. The weight of layer after layer of snow pressing down causes glaciers to freeze into ice.

(1) A only
(2) B only
(3) C only
(4) A and B only
(5) A and C only

7. Assume that the total amount of water on Earth remains constant. What would be the effect on land near the ocean if the Antarctic ice sheet began to melt?

Item 8 is based on the following diagram.

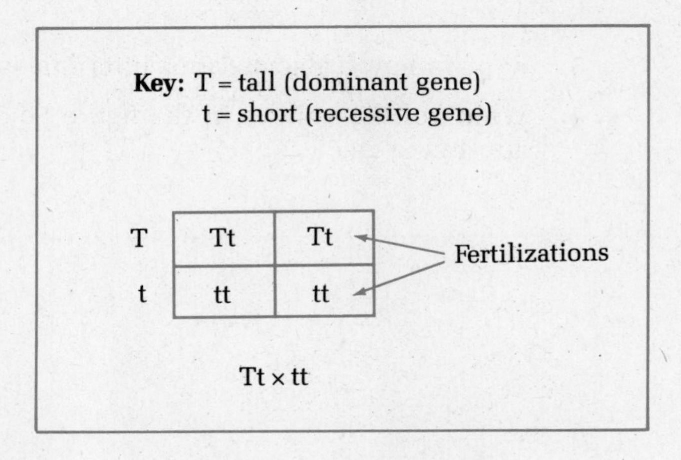

8. What chance does the cross in the figure have of producing tall offspring?

(1) $\frac{1}{1}$
(2) $\frac{1}{2}$
(3) $\frac{1}{3}$
(4) $\frac{1}{4}$
(5) $\frac{1}{5}$

---

9. An advertisement for a candy bar states that the candy
A. was chosen by more movie stars than any other brand
B. contains only 300 calories
C. is composed of the finest ingredients available
D. contains 50 percent chocolate by weight

Which of the above statements could not be proven in a chemical analysis of the candy bar?

(1) A only
(2) C only
(3) A and C only
(4) B and D only
(5) D only

Items 10 to 14 are based on the following table.

Tables of "Desirable" Weight

Height	Metropolitan Life		Gerontology Research Center Men and Women by Age (Years)				
	Men	Women	25	35	45	55	65
ft–in	*pounds*		*pounds*				
4–10	...	100–131	84–111	92–119	99–127	107–135	115–142
4–11	...	101–134	87–115	95–123	103–131	111–139	119–147
5–0	...	103–137	90–119	98–127	106–135	114–143	123–152
5–1	123–145	105–140	93–123	101–131	110–140	118–148	127–157
5–2	125–148	108–144	96–127	105–136	113–144	122–153	131–163
5–3	127–151	111–148	99–131	108–140	117–149	126–158	135–168
5–4	129–155	114–152	102–135	112–145	121–154	130–163	140–173
5–5	131–159	117–156	106–140	115–149	125–159	134–168	144–179
5–6	133–163	120–160	109–144	119–154	129–164	138–174	148–184
5–7	135–167	123–164	112–148	122–159	133–169	143–179	153–190
5–8	137–171	126–167	116–153	126–163	137–174	147–184	158–196
5–9	139–175	129–170	119–157	130–168	141–179	151–190	162–201
5–10	141–179	132–173	122–162	134–173	145–184	156–195	167–207
5–11	144–183	135–176	126–167	137–178	149–190	160–201	172–213
6–0	147–187	...	129–171	141–183	153–195	165–207	177–219
6–1	150–192	...	133–176	145–188	157–200	169–213	182–225
6–2	153–197	...	137–181	149–194	162–206	174–219	187–232
6–3	157–202	...	141–186	153–199	166–212	179–225	192–238
6–4	...	...	144–191	157–205	171–218	184–231	197–244

Heights and weights are given for people without shoes or clothing.

10. According to the Gerontology Research Center, a 25-year-old man who is 5 feet 1 inch tall should weigh no more than

(1) 93 pounds
(2) 123 pounds
(3) 127 pounds
(4) 136 pounds
(5) 145 pounds

11. A person is considered to be overweight if he or she is ten pounds or more above desirable weight. According to the Gerontology Research Center, is a 35-year-old woman who is 5 feet 5 inches tall overweight if she weighs 130 pounds?

12. Neither Metropolitan Life nor the Gerontology Research Center take into account which factor in figuring desirable weights?

(1) age
(2) bone structure
(3) height
(4) naked weight
(5) sex

13. State whether the following is a logical conclusion based on information in the Gerontology Research Center table: Because women have lighter bones and muscles, they can be relatively fatter than men the same age and still be at a "desirable" weight for their height and age.

14. Which table supports the following statement? Weight gained in early adulthood tends to be associated with more ill health than weight gained later in life.

Items 15 and 16 refer to the following passage.

From sea level upward through the atmosphere, air pressure decreases. The air also gets thinner, or less dense. This is because close to the ground, the air is pressed together by the weight of all the air above it. Higher up, the pressure is lower, so the air expands and takes up more space, cooling as it does so.

15. According to the passage, it is colder three kilometers above sea level than it is at sea level because

(1) air cools as the pressure on it decreases
(2) cold air is less dense than warm air
(3) cold air is more dense than warm air
(4) less dense air is colder than more dense air
(5) air cools as it expands

16. Water boils when water molecules gain enough heat energy to escape from the liquid into the air. The amount of energy needed for boiling depends on air pressure. Where there is lower air pressure, less energy is needed to boil water. How does air pressure affect the length of time it takes to cook eggs in boiling water at different altitudes?

(1) Boiling water is hotter at sea level than in the mountains, so cooking time is shorter.
(2) Boiling water is cooler at sea level than in the mountains, so cooking time is longer.
(3) Boiling water is the same temperature at either altitude, so cooking time is not affected.
(4) Boiling water is hotter at sea level than in the mountains, so cooking time is longer.
(5) Boiling water is cooler at sea level than in the mountains, so cooking time is shorter.

Item 17 refers to the following figure.

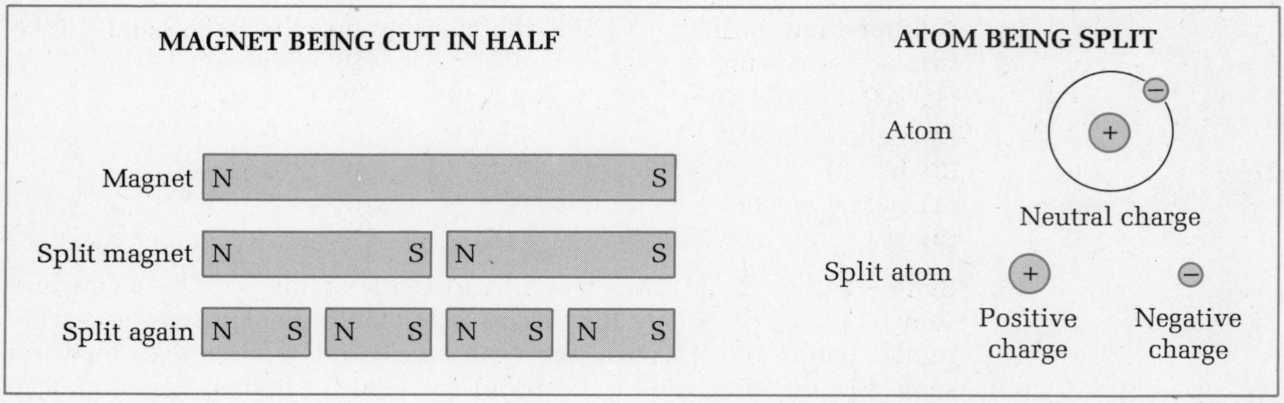

MAGNET BEING CUT IN HALF

Magnet: N S
Split magnet: N S | N S
Split again: N S | N S | N S | N S

ATOM BEING SPLIT

Atom: + −
Neutral charge
Split atom: + −
Positive charge
Negative charge

17. Electricity and magnetism are similar. A magnet has north poles and south poles, which may be compared to positive charges and negative electrical charges. A major difference, though, between electricity and magnetism is shown in the figure. What is this difference?

Items 18 to 20 refer to the following passage.

High-fiber diets have been associated with a reduced risk of human colon cancer. Nutrition scientists have come up with several explanations. One explanation is that fiber may bind to cancer-causing agents or carcinogens, taking them out of the body before they can harm the colon.

To test this theory, Swedish scientists used three chemicals known to cause cancer in the intestines of laboratory animals. Each chemical was mixed in test tubes with one of 13 different food fibers, such as oat bran and whole barley flour. The scientists discovered that all the fibers bound at least 8 percent of the chemicals. Some fibers bound 22 percent. One fiber, whole sorghum flour, bound 50 percent of the chemicals.

18. In conducting this experiment, the Swedish scientists made which of the following assumptions?

(1) Test tube conditions can be made to adequately simulate conditions in a live animal.
(2) Humans and animals get the same amount of fiber in their diets.
(3) Humans and animals get cancer from the same chemicals.
(4) Each chemical causes cancer after one exposure.
(5) Animals are less affected by cancer than are humans.

19. In the next part of the experiment, laboratory animals were fed food mixed with the different chemicals. The animal waste was then analyzed for the amount of chemical bound to food fiber. Based on the results of the first part of the study, it is logical to expect that the amount of chemical bound by fiber in the intestine would be

(1) less than in the test tube
(2) the same as in the test tube
(3) more than in the test tube
(4) different from one animal to the next
(5) unaffected by the test tube results

20. The scientists concluded that only certain food fibers protect the intestine by binding to carcinogens. Is this conclusion justified? Explain your answer.

21. Scientists at Texas Tech University discovered two crow-size skeletons about 225 million years old. The fossils, named *Protoavis*, have dinosaur-like features of clawed fingers, tails, and teeth. They also have birdlike features, such as a wishbone, a well-developed wing structure, bumps on the forearm where feathers were probably attached, and a skull like that of modern birds. Scientists know that birds first appeared on Earth about 25 million years after dinosaurs. Based on the information given, what conclusion can be drawn from the scientists' finding?

(1) Dinosaurs and birds had similar features.
(2) Protoavis is a bird.
(3) Protoavis is a dinosaur.
(4) Dinosaurs and birds evolved from a more distant common ancestor.
(5) Birds evolved directly from dinosaurs.

22. The arrangement of fossils in layers of rock helps scientists determine what kind of environment existed at a given time. Consider a site in the United States where 40-million-year-old fossils of sharks, and fish were found under the bank of a shallow creek. What kind of environment probably existed when the fossil animals were alive?

(1) a shallow marsh
(2) a lake
(3) an ocean
(4) a shallow creek
(5) a river

23. Petroleum, a mixture of simple and complex hydrocarbons, is found most often under domes of hard sandstone. There is usually natural gas just above the oil, and salt water beneath it. To explore for this valuable resource, geologists sometimes set off explosives. They then read the resulting shock waves on seismographs, the same instruments that are used to study earthquakes and the interior structure of Earth. The speed of the shock waves varies as they pass through structures of different densities.

Which of the following statements is most likely the reason why seismographs are used to find petroleum deposits?

(1) The density of rock formations is a clue to the location of deposits.
(2) The chemical composition of the surrounding rocks is unique.
(3) Earthquakes occur frequently in the vicinity of oil pools.
(4) The shock waves cause some oil leakage on the surface.
(5) The instruments are especially sensitive to organic compounds.

Items 24 to 28 refer to the following passage.

One of the ways to predict whether a chemical reaction will take place is to know the heat of formation. *Heat of formation* is the number of calories given off or absorbed when a compound is made from its elements. The amount of energy involved in formation indicates how stable a compound is formed.

In general, a compound that has a high heat of formation is very stable. Such a compound will not break down easily into its component elements. A compound that has a low heat of formation is only moderately stable and may be broken down fairly easily. A compound that has a negative heat of formation is very unstable and breaks down of its own accord.

24. If a compound has a high heat of formation, is it likely that a chemical reaction involving the compound will occur on its own?

25. NaCl has a heat of formation of 98,360 calories. Another chemical, HgO, has a heat of formation of 21,700 calories. Which is more stable?

26. A certain compound has a heat of formation of −5,926 calories. How stable would you predict it to be?

27. Chemical reactions may be either exothermic or endothermic. An exothermic reaction releases energy as it proceeds, and an endothermic reaction absorbs energy. Based on this information and that in the passage, which of the following statements are true?

(A) A compound that has a negative heat of formation is formed by an endothermic reaction.
(B) A compound that has a negative heat of formation is formed by an exothermic reaction.
(C) A compound that has a positive heat of formation is formed by an exothermic reaction.

(1) A only
(2) B only
(3) C only
(4) A and C only
(5) B and C only

28. Nitric oxide is formed from nitrogen gas and oxygen gas in automobile engines at high temperatures. Which of the following statements are true?

(A) Nitric oxide is formed in an energy-consuming reaction.
(B) Nitric oxide is formed in an energy-releasing reaction.
(C) Nitric oxide is a stable compound.
(D) Nitric oxide is an unstable compound.

(1) A only
(2) A and C only
(3) A and D only
(4) B and C only
(5) B and D only

Items 29 to 31 are based on the following figure.

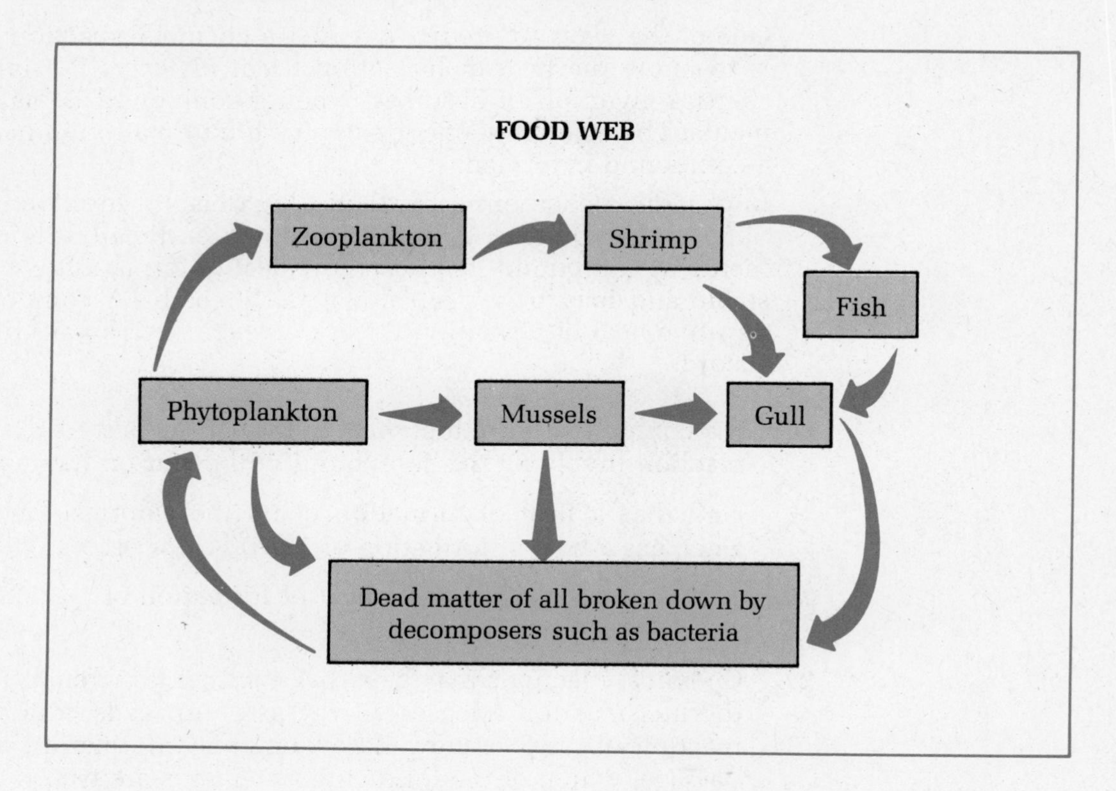

29. Which of the following members of the food web, if removed, would <u>most</u> block the free cycle of matter?

(1) bacteria
(2) phytoplankton
(3) zooplankton
(4) fish
(5) gull

30. Phytoplankton is a producer, while fish and mussels are examples of consumers in this food web. What role would a human being introduced into this food web have—consumer or producer?

31. Because of an unusually cold winter, the population of mussels goes down drastically. What effect might this have on the food web?

Items 32 and 33 are based on the following figure.

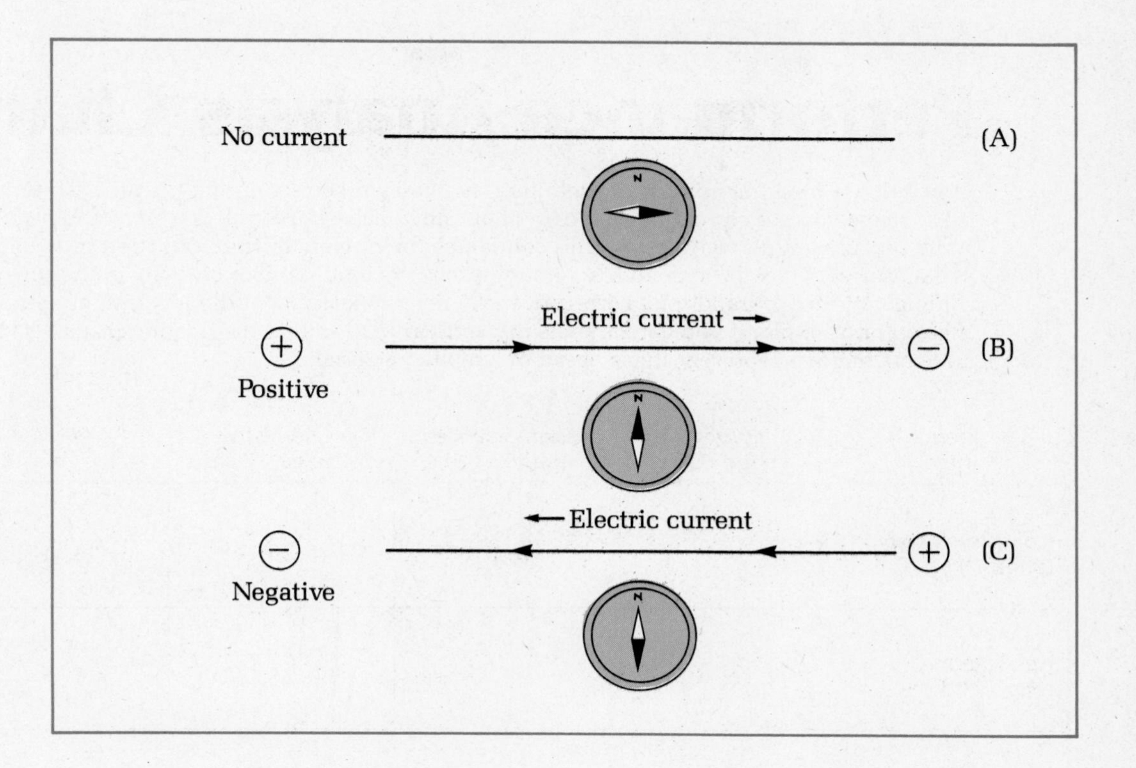

32. Which of the following produces a magnetic force field when in use?

A. bicycle
B. flashlight
C. kerosene lamp
D. toaster

(1) A and B only
(2) A and C only
(3) A and D only
(4) B and D only
(5) B, C, and D only

33. Form a hypothesis as to why the compass needle points north in (B) and south in (C).

Answers are on pages 268–269.

UNIT II TEST

Performance Analysis Chart

Directions: Circle the number of each item that you got correct on the Unit II Test. Count how many items you got correct in each row; count how many items you got correct in each column. Write the amount correct per row and column as the numerator in the fraction in the appropriate "Total Correct" box. (The denominators represent the total number of items in the row or column.) Write the grand total correct over the denominator, **33** at the lower right corner of the chart. (For example, if you got 28 items correct, write *28* so that the fraction reads *28*/**33.**) Item numbers in color represent items based on graphic material.

Item Type	*Biology (page 83)*	*Earth Science (page 106)*	*Chemistry (page 123)*	*Physics (page 142)*	TOTAL CORRECT
Comprehension (page 31)	10	3, 5	26	1, 15	**/6**
Application (page 39)	8, 11, 30	4	24, 25, 27	2, 32	**/9**
Analysis (page 46)	12, 18, 21, 22, 31	6, 7, 23	28	16, 17, 33	**/12**
Evaluation (page 57)	13, 14, 19, 20, 29		9		**/6**
TOTAL CORRECT	**/14**	**/6**	**/6**	**/7**	**/33**

The page numbers in parentheses indicate where in this book you can find the beginning of specific instruction about the various fields of science and about the types of questions you encountered on the Unit II Test.

Practice

Introduction

Most office buildings, schools, and factories conduct fire drills from time to time. During the fire drills, alarms are sounded and the people in those buildings are required to go through the safety measures that would be necessary if there were a real fire. The fire drills serve a useful purpose by making certain that everyone knows how to respond to a real fire threat. If all goes smoothly, the people can be secure knowing that they are prepared should a real emergency happen. If problems occur during the drill, they can be corrected so that they won't occur during a real crisis. The activities in this GED practice section will be your fire drill. The scores do not really count, but you can benefit a great deal from them. If you do well, you can begin to feel secure knowing that you stand a good chance of doing well on the real test. If you encounter problems, you can correct them so you will be able to achieve the best possible score at test time.

This section is filled with GED-like test questions, or *items*. It provides valuable practice on the kinds of items found on the Science test. On the pages that follow, you will find:

- Practice Items: This practice contains 66 simulated GED Test items, grouped according to the branches of science covered on the test. For example, you will find biology items grouped together, earth science items grouped together, and so on.
- Practice Test: This is a 66-item test structured like the actual Science Test. The passages are *not* grouped together according to content, but the content varies from passage to passage.

As on the actual Science Test, all the questions are multiple choice. By completing the Practice Items and the Practice Test, you will discover your strong points and weak points in science. And if you discover any weak points, *don't worry*—you will be shown how to strengthen them. The answer key not only provides the correct answer to each practice question, but it also explains *why* each answer is correct. The Performance Analysis Chart following each practice will direct you to parts of the book where you can review the skills or subjects that give you trouble.

You can use the Practice Items and the Practice Test in a number of different ways. The introductions that precede the practices will provide you with options for using them to best advantage. You may also wish to talk with your teacher to get suggestions about how best to make use of the Practice Items and Practice Test.

Practice Items

These Practice Items are similar to a real test but they are grouped according to the branches of science: biology, earth science, chemistry, and physics. As you work on the Practice Items, you will focus on one branch at a time. Your results will help you determine which skills you have mastered and which you should study further.

You can use the Practice Items in the following ways: (1) After you finish a chapter in Unit II, you can complete the section of the Practice Items that corresponds to the same branch of science. (2) You can save the Practice Items until you've completed all the chapters in Unit II. (3) You can use the Practice Items as a practice test. To do this, complete the Practice Items in one sitting. Since the actual test allows you 95 minutes, you may want to time yourself. This will give you a rough idea of how you would perform on the actual Science Test.

Compare your answers to the Practice Items to those in the answer key beginning on page 270. Whether you answer an item correctly or not, you should read the explanations of correct answers in the answer key. Doing this will reinforce your knowledge of science and develop your test-taking skills.

Regardless of how you use these Practice Items, you will gain valuable experience with GED-type questions. After scoring your work with the answer key, fill in the Performance Analysis Chart on page 200. The chart will help you determine which skills and areas you are strongest in, and direct you to parts of the book where you can review areas in which you need additional work.

PRACTICE ITEMS

Directions: *Choose the one best answer to each question.*

Biology

Items 1 to 5 refer to the following passage.

There are two types of cell division: mitosis and meiosis. Each type proceeds in a different manner and serves a different purpose. In mitosis, a cell grows to a certain size and then divides to form two cells that are exactly like the original. Mitosis allows for the growth of organisms and the repair of injured tissues. In meiosis, a cell divides twice to form four cells that differ from the original. Gametes, or sex cells for reproduction of an organism, are formed by meiosis.

Before mitosis begins, the chromosomes in the nucleus of the parent cell double, resulting in two identical sets of chromosomes. Then, in the first stage of mitosis, the centrioles outside the nucleus move to opposite sides of the cell. The nuclear membrane disappears, and fine threads form between the centrioles. These threads make up the spindle. During the second stage, the chromosomes move to the center of the cell. They attach to the spindle fibers at right angles. In the third stage, the two sets of chromosomes are pulled apart, toward the centrioles. In the final stage of mitosis, the spindle breaks apart, and a nuclear membrane forms around each set of chromosomes. The cytoplasm presses in, and a new cell membrane forms. The result is two identical daughter cells.

In meiosis, a similar process occurs. Chromosomes double, and two daughter cells are formed. However, these cells then quickly divide without any doubling of chromosomes. In meiosis, therefore, the number of chromosomes in each new cell is one-half of that in the parent cell. The complete number of chromosomes is restored when gametes join during the reproductive process.

1. A human body cell normally contains 46 chromosomes. How many chromosomes does a human sperm cell contain?

(1) 0
(2) 11.5
(3) 23
(4) 46
(5) 92

2. A scientist who uses fruit flies for research knows that a fruit-fly egg contains four chromosomes. How many chromosomes are there in each body cell in one of the parents?

(1) 2
(2) 4
(3) 6
(4) 8
(5) 10

3. A frog egg is fertilized. The unfertilized egg contained 13 chromosomes. How many chromosomes does the fertilized egg contain?

(1) 0
(2) 6.5
(3) 13
(4) 26
(5) 39

4. Which of the following statements is true when it concerns a child who closely resembles his father?

(1) The child received more chromosomes from his father than from his mother.
(2) The child received 23 chromosomes from his father and 23 chromosomes from his mother.
(3) The child did not receive a complete number of chromosomes during the reproductive process.
(4) The mother's egg contained no chromosomes.
(5) The mother's egg was the result of faulty meiosis.

5. Based on the given information, which of the following conclusions can be drawn about cell division?

(1) The daughter cells that result from mitosis will be smaller than their parent cell.
(2) Mitosis is not a continuous process.
(3) Cells that divide by meiosis do not grow larger before they divide.
(4) No spindle is formed during meiosis.
(5) Mutations in organisms occur when meoisis has not been accomplished correctly.

Items 6 and 7 refer to the following passage.

Many plant and animal activities are controlled by biological "clocks." A regular, repeating cycle is present in these activities. The pattern may repeat every few hours, every day, every month, or every year. Seasonal flights of migrating birds result from the workings of biological clocks. The opening of flower petals at daylight and their closing at dark are another example of biological clocks.

6. Based on the information given, which of the following activities is NOT controlled by a biological "clock"?

(1) the southward-bound flight of Canada geese in autumn
(2) the opening of morning-glory petals in the morning
(3) the movement of animals into areas of greater vegetation
(4) the closing of rose petals in the evening
(5) the return of swallows to their nesting spots in the spring

7. Biological clocks can be "reset." When humans travel through three or more time zones in a short period of time, they often experience a phenomenon known as "jet lag." They find that eating, sleeping, and other biological functions are disrupted. Which of the following indicates the need for resetting a biological clock?

(1) You awaken every morning at the same time.
(2) You become sleepy at about the same time every evening.
(3) You become hungry at the same times each day.
(4) You find yourself wide awake at what your watch indicates is your usual bedtime.
(5) You maintain such a high level of activity that you do not notice your hunger.

Item 8 refers to the following figure.

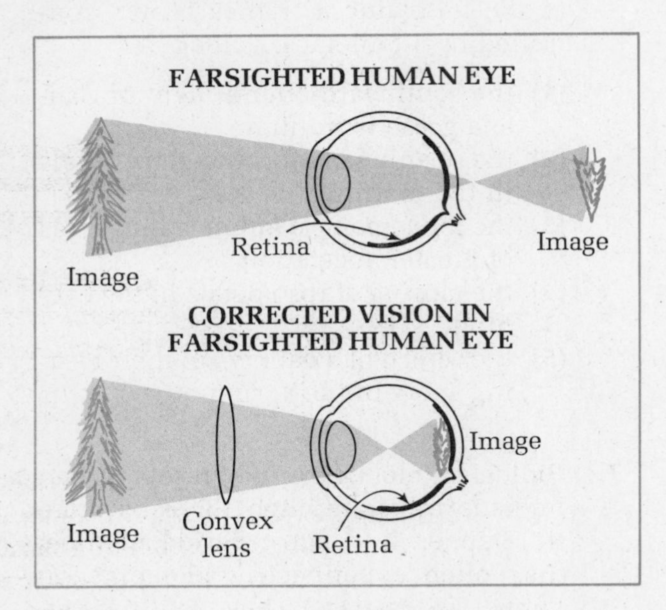

8. Farsightedness is a vision problem in which the viewer cannot see nearby objects clearly. Based on the figure, you can conclude that the biological reason for farsightedness is that

(1) the density of the eyeball makes clear reception impossible
(2) the image is focused in front of the retina
(3) the image is focused behind the retina
(4) the image is focused directly on the retina
(5) the eyeball lacks a retina

Items 9 and 10 refer to the following passage.

Throughout the roots and stems of a plant are tubes that carry the water, the food, and the minerals that are needed for the plant's life processes. Water and minerals are carried upward from roots to leaves in xylem tissue. Food that is made in the leaves by photosynthesis is carried to other parts of the plant by phloem tissue. The food-containing fluid that is carried by phloem tissue is called *sap*.

9. Some flowers secrete a sweet liquid called *nectar*. Nectar contains sugar as a result of photosynthesis. Flowers themselves do not carry out photosynthesis. Which of the following, then, best explains the presence of nectar in a flower?

(1) Rainwater that falls on the flower mixes with pollen and produces nectar.
(2) Xylem tissue carries water up to mix with sugar that is produced in the flower.
(3) Phloem tissue carries sugar and water up to the flower from parts of the plant involved in photosynthesis.
(4) Xylem tissue carries sugar and water up to the flower from parts of the plant involved in photosynthesis.
(5) Sugar seeps into the flower from the leaves that touch the petals.

10. Gnawing animals, such as rabbits and beavers, can damage a tree by chewing a circle around its trunk. Such animals chew through the phloem tissue just under the bark. Eventually, the tree dies. Which of the following describes the process the chewing animals destroyed?

(1) the movement of food from the leaves to the roots and other tissue below the cut through the xylem tissue
(2) the movement of food from the leaves to the roots and other tissue below the cut through the phloem tissue
(3) the movement of water and minerals from the roots, past the cut, to the leaves through the xylem tissue
(4) the movement of water and minerals from the roots, past the cut, to the leaves through the phloem tissue
(5) photosynthesis in the leaves of the tree

Item 11 refers to the following figure.

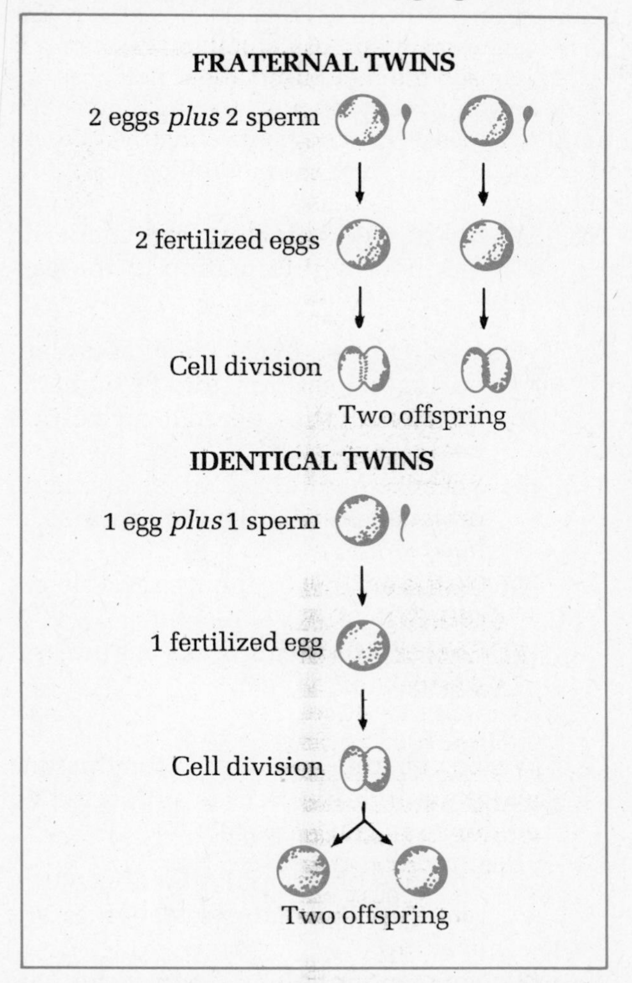

11. One of your coworkers tells you that she has a twin. From which of the following facts could you immediately determine the gender of that twin?

(1) There are older twin brothers in the same family.
(2) The coworker has a fraternal twin.
(3) The coworker has an identical twin.
(4) Both twins have black hair and brown eyes.
(5) Both twins resemble their mother more than they resemble their father.

Items 12 to 14 refer to the following passage.

Plants and animals are classified according to their similarities and degree of relatedness. The largest, most general group to which living things can belong is a kingdom. A kingdom is divided into smaller, more specific subgroups. Each subgroup is divided further into smaller and even more specific subgroups. The order of subgroups from the largest to the smallest is kingdom, phylum, class, order, family, genus, and species. The smaller the group is, the more closely related are the organisms in it. The most closely related organisms, then, are those in the same species. Organisms in the same species can breed with each other and produce fertile offspring.

More than 200 years ago, Carolus Linnaeus, a Swedish naturalist, developed a binomial, or two-name, system for the scientific naming of organisms. Linnaeus used Latin for his naming system because Latin was commonly used in the writings of scientists from many nations. In his system, the first name of an organism describes its genus; the second describes its species. *Felis tigeris* (tiger), *Felis leo* (lion), and *Felis concolor* (mountain lion) are the scientific names of some members of the cat family. These three cats belong to the same genus, but they are of different species.

12. Humans are classified as *Homo sapiens*, a description of their

(1) kingdom and family
(2) class and order
(3) family and species
(4) phylum and family
(5) genus and species

13. Which of the following situations would support the fact that the fox terrier and the beagle hound belong to the same species?

(1) You have a puppy whose mother was a fox terrier and whose father was a beagle.
(2) You find that both kinds of dogs have the same genus name.
(3) You know of an example of each kind of dog that has a long pedigree.
(4) You learn that both kinds of dogs are listed in the same phylum.
(5) You discover that each dog has two names in the name of its breed.

14. A *jackalope* is described as the offspring of an antelope and a jackrabbit. It can be logically concluded that such an animal does not actually exist because the alleged parents

(1) do not live in the same areas or eat the same foods
(2) belong to the same class of organisms
(3) have different mating seasons
(4) exist in a predator-prey relationship
(5) do not belong to the same species of organisms

---

15. A population of staphylococcal bacteria can double in size every 20 minutes.

A population of 10 staphylococci contaminates a bowl of warm soup at noon. About how many staphylococci are there in the soup by 1:00 P.M.?

(1) 20
(2) 30
(3) 40
(4) 80
(5) 160

Items 16 and 17 refer to the following passage.

Natural resources are materials supplied by nature that sustain life or that are important to economic and cultural activities. Air, soil, water, land, minerals, plants, and animals are all natural resources. Some of these resources, such as water and plants, are renewable: they are replaced over time by natural cycles. Others, such as metals and fossil fuels, are formed over millions of years and cannot be replaced.

When the population of humans on Earth was relatively small, natural resources seemed limitless. Today, however, population pressure in some areas is causing many resources to be used up at a rapid rate. For example, some species of fish have been pushed almost to extinction by overfishing. Forests in parts of the world are being destroyed to provide more land for farming. Resources such as water and air are being polluted.

To stop the destruction of natural resources, a number of practices have been developed. The wise management and use of natural resources is called *conservation*. Conservation practices include recycling, reforestation, placing limits upon hunting and fishing, and controlling the amount of pollutants that enter the environment.

16. Which of the following statements is NOT implied by information in the passage?

(1) Natural resources were abundant during the early history of humans.
(2) Many activities depend on natural resources.
(3) Because natural resources are being replaced constantly, they cannot be used up.
(4) Coal and oil are nonrenewable resources.
(5) Conservation can make natural resources last longer.

17. Which of the following conclusions about natural resources is supported by evidence in the passage?

(1) The size of a population affects the rate at which natural resources are used up.
(2) Small populations destroy the natural resources in the areas in which they live.
(3) People can move to other parts of the world when their own resources are gone.
(4) People will develop replacements for natural resources.
(5) Fish become extinct whenever they are harvested for consumption.

---

18. Soil is an important natural resource. Without soil, plants could not grow, and all other organisms that depend on plants for food could not exist. Soil is held in place by the roots of plants. When the plants are removed, the soil is eroded by water or wind. In many parts of the world, soil loss is a serious environmental problem.

Which of the following is most likely to be true of any country that is experiencing severe soil loss?

(1) Winds knock down the plants and blow away the soil.
(2) The land is steeply sloped, and the soil washes down the hill when it rains.
(3) Weather conditions or cultural practices cause large areas to be stripped of the vegetation that holds soil in place.
(4) Insects and small mammals eat young plants; as a result, the old, dying plants are never replaced.
(5) Seeds that are being sown for crops are producing plants whose roots are not strong enough to hold the soil in place.

Items 19 to 22 refer to the following passage.

A community is made up of all the organisms that live in a certain area. A population consists of all the organisms of a certain species within that area. For example, all the robins in a city block are a population; all the crabgrass plants on a lawn are a population. A community contains many populations. Usually, the two or three dominant species determine the life patterns in the community.

If a population changes, the community changes. The changing of communities is called *succession*. For example, a freshwater pond might be the first stage in a process of succession. As plant life in the pond flourishes and as dirt particles are carried into the pond, the pond eventually fills and becomes a bog. As the bog dries and firms, it becomes a meadow. In the process, one dominant species may be replaced by another. Eventually, the meadow becomes a forest. In general, if the process of succession is unhindered, a self-sustaining community will develop which will remain the same unless it is affected by such outside factors as fire or farming. This final community is called the *climax community*.

19. Which of the following does NOT describe a population?

(1) all the red clover plants in a field
(2) all the Hereford cattle on a ranch
(3) all the peach trees in an orchard
(4) all the small mammals in a forest
(5) all the humans in a large city

20. A pond was once lined by a rocky shore that had few plants and was fed by a freshwater stream. Now there are many cattails and other plants along the shore and other evidences of succession. If this process of succession continues, which of the following developments is most likely to occur?

(1) The pond will continue unchanged and will sustain the same plant and animal communities.
(2) The size of the pond will decrease, and new plants and animals will appear.
(3) The size of the pond will increase, as will the fish population.
(4) The pond will become polluted by decaying vegetation, but the same kind of plants and animals will remain in the area.
(5) The pond will become polluted by decaying vegetation, and all the animals will leave the area as a result.

21. If succession proceeds unchecked in a pond, which of the following situations would most likely occur?

(1) The pond would always exist.
(2) There would never be a climax community, for the area around the pond would be too unstable.
(3) The plant and animal life in the area would not change.
(4) Eventually, there would be a meadow where the pond is now.
(5) The pond would become and remain a bog.

22. Within the state of Ohio, there are meadows, farm ponds, hardwood forests, bogs, and cities. Which of the following hypotheses about a climax community in Ohio is the most logical?

(1) It probably is a pond that has algae, frogs, and catfish.
(2) It probably is a meadow that has goldenrod, rabbits, and cows.
(3) It probably is a bog that has cattails, snakes, and muskrats.
(4) It probably is an oak-and-hickory forest that has rabbits, foxes, and blue jays.
(5) It probably is a city that has lawns, trees, squirrels, and people.

Item 23 refers to the following figure.

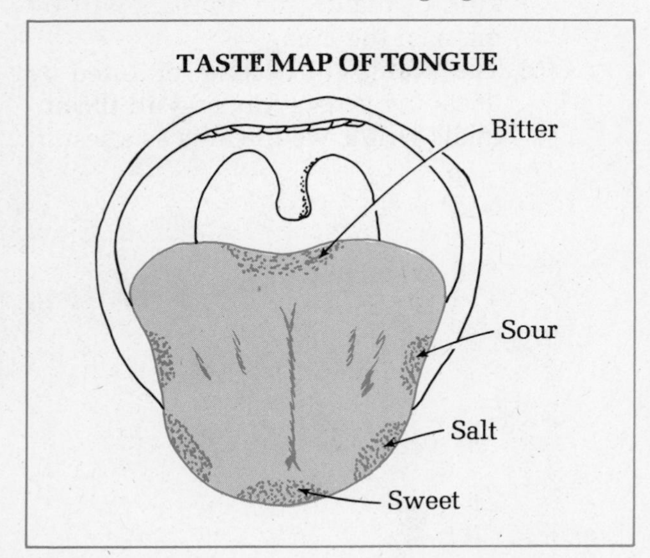

23. Based on the figure, you have the best chance of making the sweet taste of a piece of candy last by allowing the candy to melt slowly

(1) on the center of your tongue
(2) on the back of your tongue
(3) on the tip of your tongue
(4) on the forward sides of your tongue
(5) on the rear sides of your tongue

---

24. The seeds that are produced by many different kinds of plants are useful to people. Some seeds can be planted to produce new crops, or they can be eaten directly as food. For example, corn, rice, and wheat plants produce seeds that can be planted or eaten. In parts of the world where food supplies are very limited, people have sometimes eaten their entire supply of seeds instead of planting them.

Which of the following effects is most likely to result from this action?

(1) The food shortage will worsen because no new crops can be grown.
(2) The food shortage will spread to countries that export seeds.
(3) The food shortage will be solved because people will use the seeds for food.
(4) The food shortage will be solved because people will turn to eating animals when the plants are gone.
(5) The food shortage will be solved because people will develop new sources of food.

Item 25 refers to the following equations.

$$6CO_2 \text{ (carbon dioxide)} + 6H_2O \text{ (water)} \xrightarrow[\text{chlorophyll}]{\text{light}} C_6H_{12}O_6 \text{ (glucose)} + 6O_2\uparrow \text{ (oxygen)}$$

PHOTOSYNTHESIS

$$C_6H_{12}O_6 + 6O_2 \longrightarrow 6CO_2\uparrow + 6H_2O + \textit{energy}$$

RESPIRATION

25. Which of the following conclusions is justified by the information in the equation?

(1) Plants use the same molecules of carbon dioxide and water over and over.
(2) The energy that is produced by respiration is the same energy that is used in photosynthesis.
(3) The raw materials of photosynthesis are the waste products of respiration.
(4) Photosynthesis and respiration occur simultaneously.
(5) Photosynthesis and respiration are unrelated processes.

26. A cement wall that borders a freeway was constructed in order to reduce noise and air pollution for people living nearby. The cement wall proved an ineffective solution. Some time later, the area between the freeway and the cement wall was planted with pine trees.

Which of the following statements gives the best justification for the planting of the pine trees?

(1) The pine trees shade the cement wall from sunlight so that the cement will not age as quickly as it would if trees were present.
(2) The pine needles help to trap noise; trees take in carbon dioxide and carbon monoxide and give off oxygen.
(3) The pine trees protect the cement wall from wind erosion.
(4) The pine roots protect the soil from the effects of acid rain.
(5) The pine trees absorb vibrations from heavy traffic and thus prevent the vibrations from cracking both the wall and nearby buildings.

Item 27 refers to the following figure.

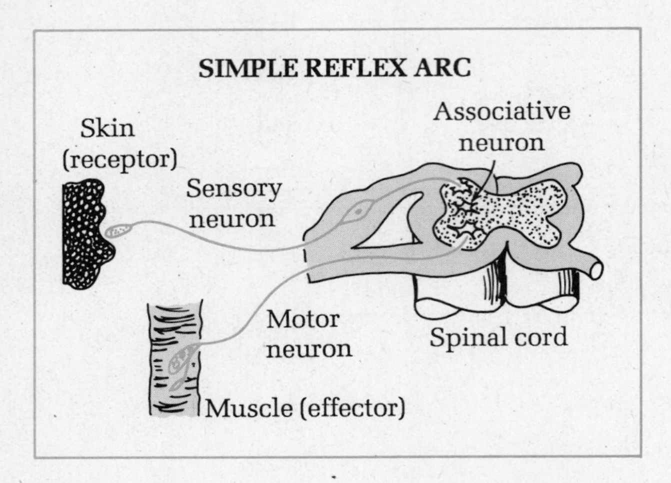

27. Which of the following statements can be assumed, based on information in the figure?

(1) The spinal cord is not part of a reflex arc.
(2) The brain is not part of a reflex arc.
(3) Nerves are not involved in a reflex arc.
(4) Simple reflexes are inherited.
(5) Simple reflex actions take place more rapidly than do conditioned reflex actions.

Items 28 and 29 refer to the following passage.

A *mutation* is a random change in the gene or the chromosome of an organism. Some mutations result in visible changes; others result in changes that are hardly noticeable. In either case, such changes may be passed on to the offspring of the organism. Some mutations are beneficial—they increase an organism's ability to survive in its environment. Most mutations are harmful—they reduce an organism's ability to compete with others of its species.

28. Which of the following describes a mutation that would help an organism to survive?

(1) a wingless grasshopper
(2) an albino field mouse
(3) a short-necked giraffe
(4) a desert plant that has shallow, branching roots
(5) a lion with no tail

29. If a mutation is harmful, which of the following generalizations has the highest probability of being proven true?

(1) The organism will survive to produce many offspring that are exactly like itself.
(2) The mutation will reverse itself in the next generation.
(3) The organism will not survive.
(4) The organism will have an advantage over the other members of its species.
(5) The organism will pass the mutation to other members of its species.

Item 30 refers to the following figure.

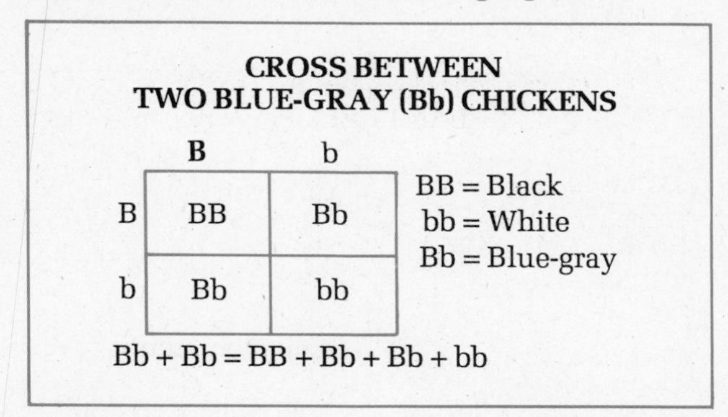

30. A breed of chicken comes in three colors—black, white, and blue-gray, which is a hybrid that results from a cross between black and white. Blue-gray chickens, however, do not breed true, as illustrated in the figure. Which of the following matings would probably produce the greatest number of blue-gray chickens?

(1) a white chicken with a white chicken
(2) a white chicken with a blue-gray chicken
(3) a black chicken with a black chicken
(4) a black chicken with a blue-gray chicken
(5) a black chicken with a white chicken

Items 31 and 32 refer to the following passage.

Lichens are complex organisms that grow on trees and on rocks. Lichens consist of two kinds of organisms—algae and fungi—that live together in a symbiotic relationship. The photosynthetic algae use chlorophyll to make food, which is used by the fungi. The fungi supply water, which keeps the algae from drying out.

31. If the description above is a typical illustration, which of the following generalizations is most likely to be true about symbiotic relationships?

(1) Symbiosis is not a continuous relationship; it is frequently interrupted by outside factors.
(2) Both organisms in such a relationship benefit from the symbiosis.
(3) In a symbiotic relationship, one organism benefits at the expense of the other.
(4) Symbiotic relationships can exist between two plants, between two animals, or between a plant and an animal.
(5) Symbiotic relationships can exist only between very simple kinds of organisms.

32. Which of the following statements about lichens is supported by evidence in the passage?

(1) A lichen's fungi can exist apart from a symbiotic relationship with algae.
(2) Lichen can exist only on the surfaces of rocks and trees.
(3) Lichen help to make soil by breaking up rocks into increasingly small particles.
(4) If the lichen's algae die, the fungi will reproduce in greater numbers.
(5) If the lichen's algae die, the fungi will die as well.

Item 33 refers to the following figure.

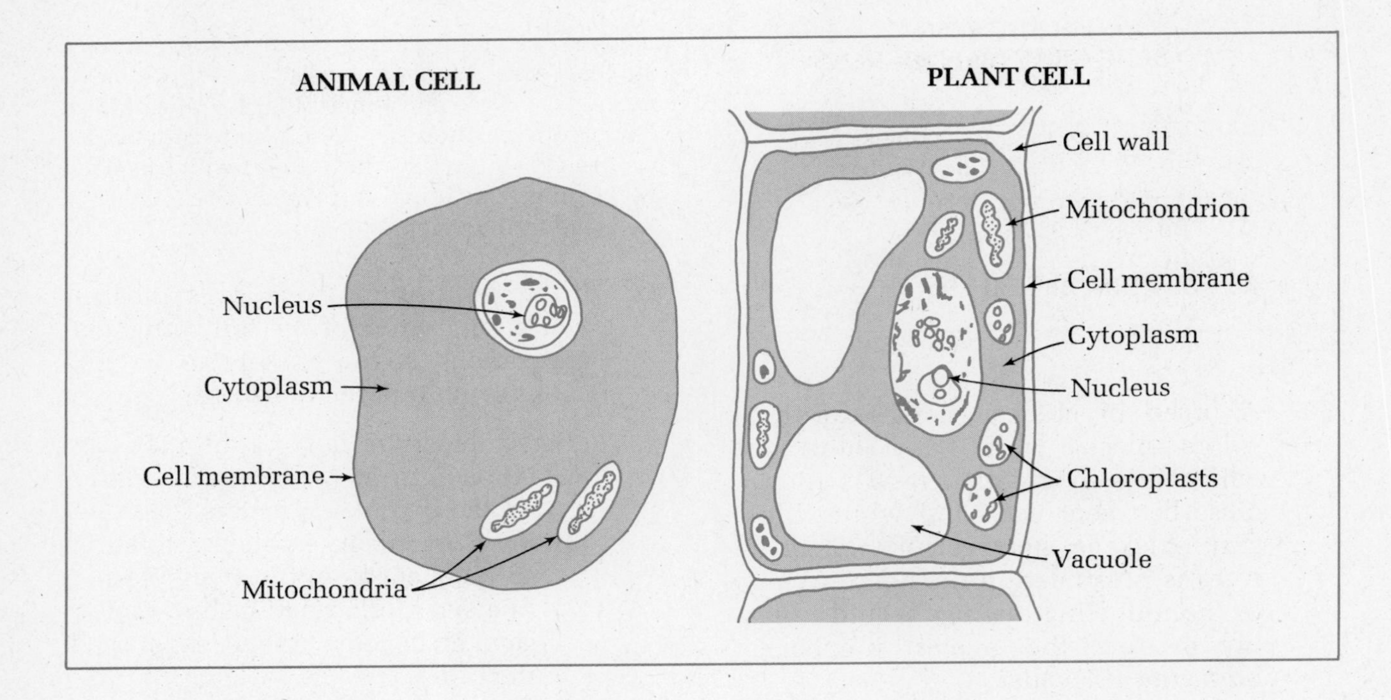

33. Based on the figure, which of the following structures are found only in plant cells?

(1) nucleus and cytoplasm
(2) nucleus and cell membrane
(3) mitochondrion and nuclear membrane
(4) cell membrane and mitochondrion
(5) cell wall and chloroplasts

Earth Science

Items 34 to 37 refer to the following figure.

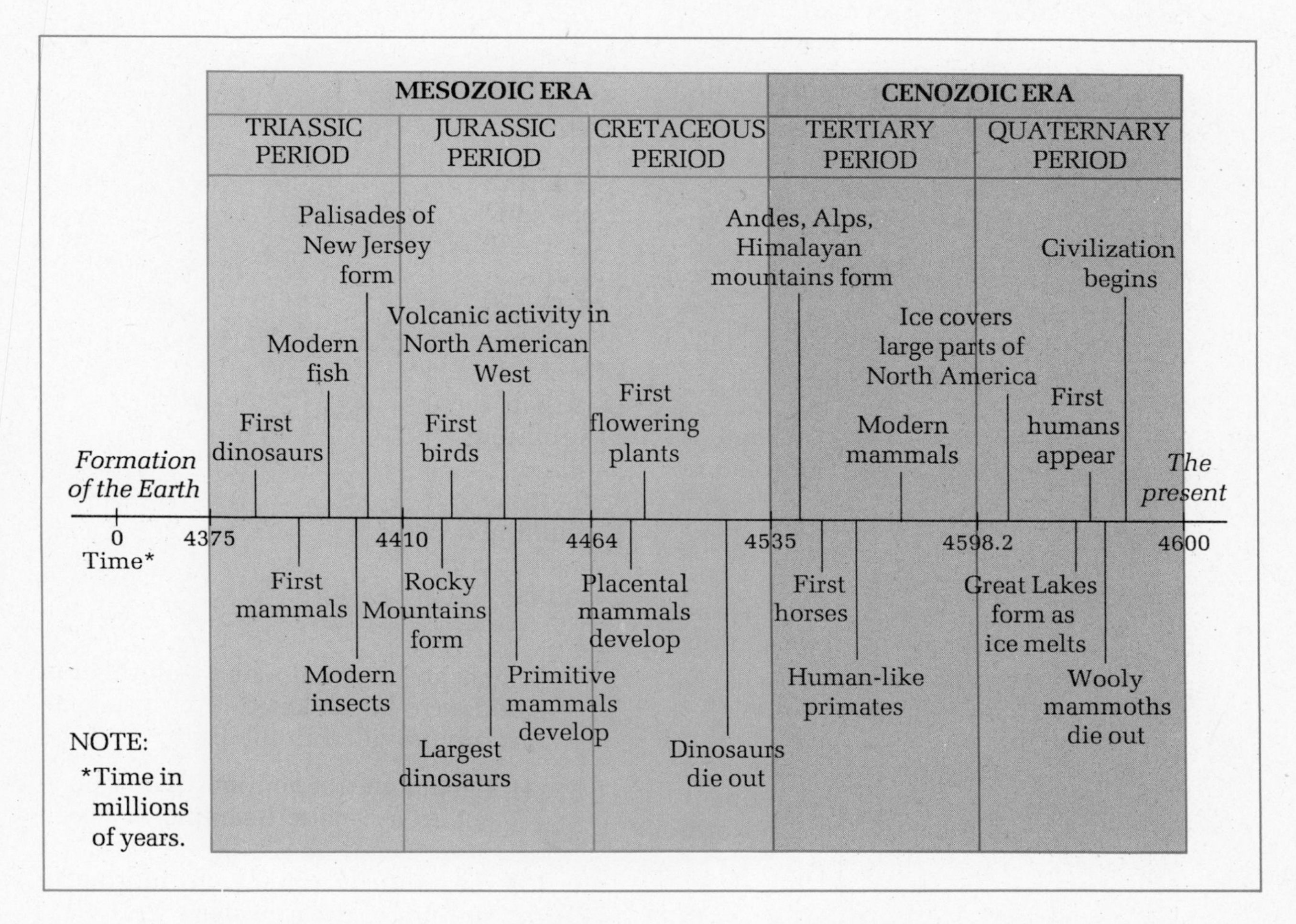

An era is the largest division of geological time. Eras are divided into periods. This time line shows the two most recent eras in the history of Earth.

34. Which of the following events best characterizes the Mesozoic Era?

(1) the freezing and the melting of large ice masses
(2) the appearance and the extinction of wooly mammoths
(3) the appearance and the development of primates
(4) the appearance and the development of mammals
(5) the beginning of civilization

35. Which of the following source materials would be the LEAST helpful for someone who is studying the extinction of dinosaurs?

(1) an encyclopedia article about the Cretaceous Period
(2) a magazine article entitled "A History of Earth: The Past Million Years"
(3) a museum display of fossils that date from the time of the first mammals
(4) a book entitled *Earth at Age 4.4 Billion*
(5) in a book, a chapter that compares present-day animals to animals of the Cretaceous Period

36. In reviewing the natural-history displays at a museum, the curator discovers that one of the items has been labeled incorrectly. Which of the following labels should be changed?

(1) a horse bone that is labeled *150 million years old*
(2) a dinosaur footprint that is labeled *250 million years old*
(3) a fish fossil that is labeled *200 million years old*
(4) a primate skull that is labeled *50 million years old*
(5) a wooly-mammoth fossil that is labeled *1 million years old*

37. Which of the following statements could be disproved by the information that is provided?

(1) Wooly mammoths existed during the Quaternary Period.
(2) The appearance and the extinction of dinosaurs signaled the beginning and the end of an era.
(3) The Andes are the oldest mountains in the world.
(4) Primitive mammals and dinosaurs existed on Earth at the same time.
(5) Human beings did not live in North America during the Ice Age.

Item 38 refers to the following list.

Composition of Ocean Water

Element	*Percent by Weight*
oxygen, hydrogen	96.5
chlorine	1.9
sodium	1.1
magnesium, sulfur, calcium, potassium, bromide, carbon, strontium, silicon, fluorine, aluminum, phosphorus, iodine	0.5

38. Which of the following information could NOT be obtained from the data that are provided in the list?

(1) the amount of sodium that is present in a 5-pound sample of ocean water
(2) the ratio by weight of chlorine to sodium in ocean water
(3) the elements that make up more than 95% of ocean water
(4) the amount of magnesium that is present in a 5-pound sample of ocean water
(5) the likelihood of finding more chlorine or more fluorine in a 5-pound sample of ocean water

Items 39 to 43 refer to the following information.

Minerals can be identified by certain physical properties. Below are five of these properties.

(1) **luster**—the way in which a mineral reflects light from its surface
(2) **hardness**—the ability of a mineral to resist being scratched
(3) **streak**—the color of the powder that is left by a mineral when it is rubbed against a hard, rough surface
(4) **density**—the mass-per-unit volume of a sample of a mineral
(5) **cleavage or fracture**—the tendency of a mineral to break evenly along definite lines (cleavage) or to break unevenly (fracture)

Each of the following items uses one of the five properties that are defined above to describe the identification of minerals. For each item, choose the one category that best describes the properties. Any of the categories above may be used more than once in the following set of items.

39. When talc is rubbed against a piece of unglazed porcelain, a white mark is left. When graphite is rubbed against the porcelain, a black mark is left. Which property of porcelain makes it a useful tool for identifying minerals that streak?

(1) luster
(2) hardness
(3) streak
(4) density
(5) cleavage or fracture

40. A sample of mineral *X* sinks in water, but a same-size sample of mineral *Y* floats. Which property distinguishes mineral *X* from mineral *Y*?

(1) luster
(2) hardness
(3) streak
(4) density
(5) cleavage or fracture

41. Nuggets of pyrite, often called "fool's gold," look like gold because they are shiny and metallic. Which property causes pyrite to be mistaken for gold?

(1) luster
(2) hardness
(3) streak
(4) density
(5) cleavage or fracture

42. When a hammer and a chisel are applied to a sample of mica, the mica can be removed in layer after layer of very thin sheets. Which property is illustrated by this phenomenon?

(1) luster
(2) hardness
(3) streak
(4) density
(5) cleavage or fracture

43. The Mohs Scale ranks calcite above gypsum because it can scratch gypsum, but it ranks calcite below fluorite because it cannot scratch fluorite. Which property forms the basis for the Mohs Scale?

(1) luster
(2) hardness
(3) streak
(4) density
(5) cleavage or fracture

44. Fossils form when living things die and are buried in sediments that harden into rock. Very few organisms leave fossils. The soft parts of an organism tend to decay or to be eaten by animals before a fossil forms. Most fossils are found in areas that were once under water; in such places, sediments of sand and mud buried dead organisms quickly. According to the passage, which of the following factors has the greatest influence over the failure of a dead organism to become a fossil?

(1) lack of available sediments
(2) lack of water
(3) being buried too quickly
(4) decaying or being eaten too quickly
(5) being buried in rock that hardens too quickly

45. Sedimentary rocks form near the surface of Earth when sediments of sand, mud, and organic remains harden. Igneous rocks form deep in the crust or mantle of Earth when molten rock hardens into crystals. Metamorphic rocks, which also are formed deep in Earth, result when intense heat and pressure change existing rocks.

Which of the following statements is implied by the information in the passage?

(1) Pressure and temperature increase greatly as one moves deeper into Earth.
(2) Conditions deep in Earth are not much different from conditions near the surface.
(3) Pressure near the surface of Earth is greater than pressure near its mantle.
(4) Fewer kinds of rocks will be found near the surface of Earth than deep within Earth's crust.
(5) Rocks found near the surface of Earth are harder than rocks found near its mantle.

Chemistry

46. Two factors—temperature and pressure—affect the amount of gas that will dissolve in a liquid. The amount of gas that will dissolve in a liquid increases as temperature decreases and pressure increases.

The cap on a bottle of soda is loosened, and the bottle is left standing by a sunny window. After several hours, which of the following phenomena will have occurred?

(1) The bottle of soda will have exploded.
(2) The soda will be unchanged.
(3) The "fizz" in the soda will have increased.
(4) The "fizz" in the soda will have decreased.
(5) Additional gases from the air will have dissolved in the soda.

Items 47 to 50 refer to the following passage.

Chemical compounds are combinations of elements. The properties of a compound are usually different from those of the elements that make it up. Two common types of compounds are acids and bases.

Acids contain hydrogen. In solution, many acids separate to form ions (charged atoms) of hydrogen and other atoms. In concentrated form, acids can eat through living tissue and even metal. In amounts controlled by the body, hydrochloric acid (HCl) in your stomach breaks down any meat you eat. Excess hydrochloric acid eats away the stomach lining, causing an ulcer. Both hydrochloric acid and the sulfuric acid (H_2SO_4) used in car batteries can break down metal. Acids have a sour taste. Citric acid (which is found in lemon juice) and acetic acid (which is found in vinegar) are dilute acids.

Dyes that are called *indicators* change color in acids and bases. Litmus paper, for example, turns blue in bases and pink in acids.

The properties of bases are often just the opposite of acids. Instead of hydrogen ions, bases have hydroxide ions (OH^-) in solution. The common base sodium hydroxide (NaOH) is found in household lye.

When an acid and a base react chemically, they form salts, another kind of chemical compound. For example, HCl and NaOH react to form water and the salt NaCl. Although both HCl and NaOH are strong chemicals that cause burns if touched, they form simple saltwater by reacting chemically.

47. It is true that acids and bases

(1) are both compounds
(2) can both be found on a periodic table of elements
(3) are both salts
(4) both carry a positive charge
(5) both carry a negative charge

48. Hydrochloric acid (HCl) reacts with the base potassium hydroxide (KOH). Which of the following compounds is one product of this reaction?

(1) HCl
(2) H_2SO_4
(3) KCl
(4) KOH
(5) $NaSO_4$

49. A substance that contaminates a home water supply eats through copper pipes, turns litmus paper blue, and reacts with HCl. It probably is

(1) a salt
(2) a solution
(3) a concentrated acid
(4) a dilute acid
(5) a base

50. Hydrochloric acid (HCl) and sodium hydroxide (NaOH) react in a one-to-one ratio of molecules to form saltwater. Forty million molecules of HCl are mixed with thirty million molecules of NaOH in water. After the reaction is complete, is the water safe to drink?

(1) No; excess base is left.
(2) No; excess acid is left.
(3) Yes; the acid and the base are now gone.
(4) Yes; only saltwater remains.
(5) Yes; the solution is neutralized.

Items 51 to 53 refer to the following table.

Specific Heats of Common Substances

Material	Specific Heat (cal/gram-C°)
alcohol	0.594
aluminum	0.22
brass	0.09
carbon	0.170
copper	0.092
glass	0.20
gold	0.030
ice	0.50
iron	0.107
lead	0.030
silver	0.056
steam	0.5
water	1.0
zinc	0.092

The specific heat of a substance is the amount of heat that is required to raise the temperature of 1 gram of that substance 1° Celsius.

51. Compared to a gram of water, a gram of ice will increase in temperature

(1) one-fourth as fast
(2) half as fast
(3) at the same rate
(4) twice as fast
(5) five times as fast

52. Steam enters a turbine at 400°C and leaves at 165°C. How much heat does each gram of steam give up to the turbine as it passes through?

(1) 0.5 calories
(2) 117.5 calories
(3) 165 calories
(4) 235 calories
(5) 400 calories

53. Iron block A is heated until its temperature increases 50°C. Iron block B, which receives the same amount of heat, increases 150°C. Which of the following statements must be true about iron blocks A and B?

(1) The mass of block A is one-third the mass of block B.
(2) The mass of block A is one-half the mass of block B.
(3) The mass of block A is the same as the mass of block B.
(4) The mass of block A is twice the mass of block B.
(5) The mass of block A is three times the mass of block B.

---

54. A *solid* is any form of matter that has definite shape and definite volume. A *liquid* is any form of matter that has definite volume but not definite shape. A *gas* is any form of matter that has neither definite shape nor definite volume.

Which of the following conclusions can be supported by the above information?

(1) Milk that fills a pint jar also can fill a gallon jug.
(2) Perfume vapors that occupy 4 cubic centimeters in a perfume bottle also can fill a room that measures 60 cubic meters.
(3) A piece of wood that measures 5 centimeters × 25 centimeters by 1 centimeter will fit exactly into a square space in an ice cube tray that measures 125 cubic centimeters.
(4) Helium that fills a 1-liter balloon will occupy the same amount of space if released into a 50-liter tank.
(5) Cake that is made from batter which is poured into a round pan before baking will be the same shape as cake that is made from dough which is poured into a square pan.

55. Chemical reactions involve the breaking and the forming of bonds between atoms. Breaking bonds requires energy; forming bonds releases energy.

Compounds AX and BY react to form compounds AY and BX:

$$AX + BY \longrightarrow AY + BX$$

Energy in the form of heat is released by the reaction.

According to the information that is provided, which of the following statements explains why heat is released?

(1) The energy that is required to break the bonds A-X and B-Y is greater than the energy that is required to break the bonds A-Y and B-X.
(2) The energy that is required to break the bonds A-X and B-Y is greater than the energy that is released by the formation of bonds A-Y and B-X.
(3) The energy that is required to break the bonds A-X and B-Y is less than the energy that is released by the formation of bonds A-Y and B-X.
(4) The breaking of bonds A-X and B-Y requires no energy, but the formation of bonds A-Y and B-X releases energy.
(5) The formation of bonds A-Y and B-X requires no energy, but the breaking of bonds A-X and B-Y releases energy.

Physics

56. *Energy* is described as the ability to do work. The two most common forms of energy are potential energy and kinetic energy. *Potential energy* is the ability of an object at rest to create force, as in the case of a coiled spring. *Kinetic energy* is the energy that an object has when it is in motion, as in the case of a spring that flexes.

Based on the definitions in this passage, which of the following can you assume is a valid generalization about the energy sequence that takes place when work is performed?

(1) kinetic energy → potential energy → work
(2) potential energy → work → potential energy
(3) potential energy → kinetic energy → work
(4) potential energy → work → kinetic energy
(5) work → kinetic energy → potential energy

57. Newton's second law of motion states that an object accelerates because a force acts upon it. The rate of the force (F) depends upon the mass (m) of the object and the size of the acceleration (a). This law is summarized by the formula $F = ma$.

A bowler who normally uses an 8-pound bowling ball picks up a 12-pound ball and rolls it with the same amount of force that she normally uses. Which of the following situations is most likely to result?

(1) The ball rolls down the alley the same as usual.
(2) The ball gains speed about $1\frac{1}{2}$ times as fast as usual.
(3) The ball gains speed about $\frac{2}{3}$ as fast as usual.
(4) The ball gains speed about $\frac{1}{4}$ as fast as usual.
(5) The ball does not travel down the alley at all.

Item 58 refers to the following figure.

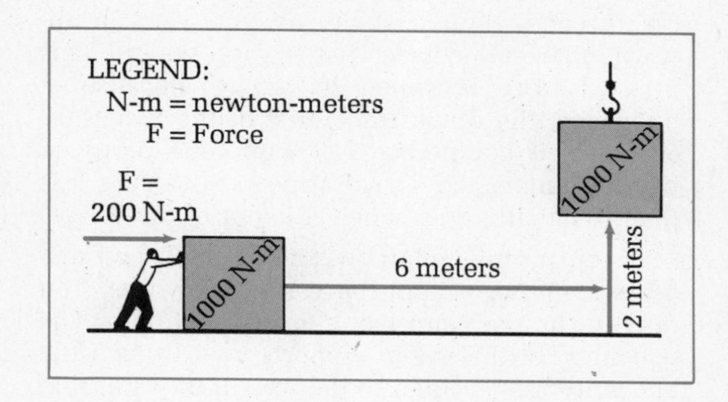

The work done on an object is equal to the product of the force that is acting on the object times the distance that the object is moved ($W = F \times D$).

58. In the figure above, a worker pushes, with a force of 200 newtons, a box along a flat surface. Then the box is lifted by a crane. Disregarding the effects of friction, which of the following represents the total amount of work that was done on the box?

(1) 800 newton-meters
(2) 1200 newton-meters
(3) 2000 newton-meters
(4) 3200 newton-meters
(5) 9600 newton-meters

Items 59 to 62 refer to the following passage.

Why does an object float? *Gravity* is a force that draws objects down, toward the center of Earth. The weight of an object is the result of this attraction. Gravity would cause an oil tanker, for example, to weigh many tons; yet, an oil tanker at sea floats rather than sinks. This phenomenon occurs because liquids, such as water, exert an upward force on the object that opposes gravity. This upward force is called *buoyancy*. Because of buoyancy, the downward force of the weight of an object is counteracted. Therefore, a partly or wholly submerged object appears to weigh less than it actually does when it is out of the water.

If you lie in a small swimming pool, you displace some water and the water level rises. You float in the pool and feel almost weightless. The buoyancy of the water supports you. In fact, the buoyant force is equal to the weight of water that you displace. Therefore, an object floats when its weight is equal to the weight of the displaced fluid. An object sinks only until it displaces enough fluid to equal its own weight. A 10-ton ship floats by sinking far enough to displace 10 tons of water. The buoyant force cancels the downward weight.

If an object has a large volume, it can displace more water than can an object that has a small volume. For example, a solid 1-pound iron ball sinks, but a large 1-pound beach ball floats.

59. An ice cube floats with its top edge slightly above the water surface in a glass. The weight of the ice cube is equal to

(1) the density of the air around it
(2) the force of gravity that is pulling on it
(3) the density of the water in the glass
(4) the weight of the water in the glass
(5) the weight of the water that it displaces

60. A penny is dropped into a glass of water and promptly sinks to the bottom. Why does it sink rather than float?

(1) It cannot displace enough water to equal its weight.
(2) Its weight is less than that of the water around it.
(3) Its weight is equal to that of the water around it.
(4) Its volume is too great in proportion to its weight.
(5) Its flatness makes it insupportable by the amount of water in the glass.

61. An overweight person is advised by a physician to exercise in water rather than on land in order to reduce stress on his or her joints. Which of the following is most likely the reason for this advice?

(1) Air has more buoyancy than water.
(2) Gravity exerts a greater force on land than in the water.
(3) Such a person displaces a greater volume of water than air.
(4) Such a person displaces enough water to equal his or her weight.
(5) The buoyancy of water can support greater weight than can air.

62. A ship is being designed for greatest efficiency in flotation. The following design features are being considered.

A. small B. huge C. hollow
D. solid E. streamlined

Which combination of features would promise the greatest efficiency in flotation?

(1) B and C
(2) B and D
(3) A and C
(4) A and D
(5) A and E

Items 63 to 66 refer to the following information.

The motion of an object can be described in a number of ways. Below are five quantities that can describe a moving object.

(1) **distance (d)**—the length of the path that is traveled by an object
(2) **displacement (s)**—the length of the line between the object's starting position and its final position
(3) **velocity (v)**—the distance that is traveled per unit of time $\left(v = \frac{d}{t}\right)$
(4) **acceleration (a)**—the change in velocity per unit of time $\left(a = \frac{v}{t}\right)$
(5) **momentum (mv)**—the product of an object's mass times its velocity

Each of the following items describes motion with reference to one of the five qualities that are defined above. For each item, choose the one category that best describes the motion. Any of the categories above may be used more than once in the following set of items.

63. A man leaves his house, walks four blocks north to buy a newspaper and then walks two blocks east to the post office. Next, he walks 1 mile west to have dinner with a friend. Which quantity describes how far this man is from his house?

(1) distance
(2) displacement
(3) velocity
(4) acceleration
(5) momentum

64. A subcompact car and a heavy truck leave a rest stop at the same time. Both vehicles travel 2 miles in the first 5 minutes, 4 miles in the next 5 minutes, and 6 miles in the last 5 minutes before stopping at a red light. Which quantity was NOT the same for both the car and the truck?

(1) distance
(2) displacement
(3) velocity
(4) acceleration
(5) momentum

65. Two friends start out together in a race. Friend 1 covers $\frac{1}{2}$ mile in the first 3 minutes but then loses his way and must cover an extra $\frac{1}{4}$ mile at the same speed to get back on course. Then he covers the next $\frac{1}{2}$ mile in 2 minutes. Friend 2 stays on course, covers $\frac{1}{8}$ of a mile every minute, and joins friend 1 at the finish line. Which quantity was the same for both friends?

(1) distance
(2) displacement
(3) velocity
(4) acceleration
(5) momentum

66. A 1-pound hockey puck that is already in motion skids from Point A to Point B in 2 seconds, slides from Point B to Point C in 4 seconds, and then continues to slide. Point A is 5 yards from Point B; Point C is 10 yards from Point B. Which quantity has a value of zero (0) as the puck moves from Point A to Point C?

(1) distance
(2) displacement
(3) velocity
(4) acceleration
(5) momentum

Answers are on pages 270–272.

PRACTICE ITEMS

Performance Analysis Chart

Directions: Circle the number of each item that you got correct in the Practice Items. Count how many items you got correct in each row; count how many items you got correct in each column. Write the amount correct per row and column as the numerator in the fraction in the appropriate "Total Correct" box. (The denominators represent the total number of items in the row or column.) Write the grand total correct over the denominator **66** at the lower right corner of the chart. (For example, if you got 56 items correct, write *56* so that the fraction reads *56*/**66**.) Item numbers in color represent items based on graphic material.

Item Type	*Biology (page 83)*	*Earth Science (page 106)*	*Chemistry (page 123)*	*Physics (page 142)*	TOTAL CORRECT
Comprehension (page 31)	6, 7, 8, 12, 16, 19, 30, 33	34, 44	47, 51	58, 59	**/14**
Application (page 39)	1, 2, 3 13, 14, 15, 20, 23, 28	36, 39, 40 41, 42, 43	48	60, 63, 64, 65 66	**/21**
Analysis (page 46)	4, 9, 10, 11, 18, 21, 24, 25, 27,	35, 38, 45	46, 49, 52, 53, 55	57, 61	**/19**
Evaluation (page 57)	5, 17, 22, 26, 29, 31, 32	37	50, 54	56, 62	**/12**
TOTAL CORRECT	**/33**	**/12**	**/10**	**/11**	**/66**

The page numbers in parentheses indicate where in this book you can find the beginning of specific instruction about the various fields of science and about the types of questions you encountered on the Practice Items.

Practice Test

As in the actual Science Test, the items on the following Practice Test appear in a mixed order. On the actual test, after completing items from one branch of science, you will be required to "switch gears" and answer questions from a different branch. This Practice Test is structured in the same way. It is also the same length (66 items) as the actual test and as challenging. By taking the Practice Test, you can gain valuable test-taking experience and will know what to expect when you sit down to take the actual Science Test.

You can use the Practice Test in the following ways: (1) To get hands-on, test-taking experience, you may wish to take the Practice Test under conditions similar to those of the actual test. To do this, complete the Practice Test in one sitting and try to answer all the questions within a 95-minute time limit. (2) If you want, you can take the practice test in sections. For example, you can plan to answer ten items a day or complete a third of the test at a time. Although this does not simulate the actual testing situation, your results will still give you a fairly good idea of how well you would do on the real test.

Compare your answers to the Practice Test to those in the answer key beginning on page 273. Whether you answer an item correctly or not, you should read the information that explains each correct answer. This will help you reinforce your knowledge of science and enhance your test-taking skills.

However you decide to take the Practice Test, your score will point out your strengths and weaknesses in science. The Performance Analysis Chart at the end of the test will help you identify those strengths and weaknesses and direct you to parts of the book where you can review specific topics.

PRACTICE TEST

Directions: *Choose the one best answer to each question.*

Item 1 refers to the following passage.

The force that is exerted on an object by gravity depends on the mass of an object (the amount of matter) and its distance from the body that is pulling it. The weight of an object depends on its mass and on the pull of gravity. If either factor increases, the weight of the object will increase.

1. According to the passage above, which of the following statements best explains the weightlessness of astronauts in outer space?

(1) They have no mass in space.
(2) They move too fast to feel any weight.
(3) They are too far from Earth to feel its gravitational pull.
(4) They are closer to the moon than to Earth, and the moon has no gravity.
(5) They gain mass as they move farther from Earth.

---

2. Earth is a sphere that is made of three layers. The outer layer, the crust, is made of lightweight rock, which is formed into mountains, valleys, and plains. Bodies of water are found on the crust. Inside the crust and extending down about 1800 miles is the hot, molten mantle. The dense core beneath the mantle extends another 2000 miles to the center of Earth.

According to this passage, in which layer would you expect to find living organisms?

(1) the center of Earth
(2) the core
(3) the mantle
(4) the crust
(5) upper atmosphere

Items 3 to 7 are based on the following passage.

Immunity is an individual's ability to resist infection by harmful agents of disease. One way in which the body protects itself against these invaders is through the use of antibodies, which combat foreign bodies in the bloodstream and guard against future infection by the same agent. Listed below are five categories of immunity.

(1) **Active immunity—naturally acquired:** An individual is naturally exposed to a disease in the environment and produces antibodies in response.

(2) **Active immunity—artificially acquired:** A weak or a dead form of a disease-causing agent is injected into an individual whose body produces antibodies in response.

(3) **Passive immunity—naturally acquired:** Antibodies are passed from mother to fetus or infant.

(4) **Passive immunity—artificially acquired:** Antibodies themselves are injected into an individual.

(5) **Susceptibility:** An individual does not have or does not acquire immunity to a disease-causing agent.

Each of the following items describes a biological condition that involves one of the categories above. For each item, choose the one category that best describes the given situation. The categories may be used more than once in the set of items, but no one question has more than one best answer.

3. A young girl whose brother had chicken pox last year is a class member at a school where all the children except her come down with the disease. Which type of immunity does the young girl have?

(1) active immunity—naturally acquired
(2) active immunity—artificially acquired
(3) passive immunity—naturally acquired
(4) passive immunity—artificially acquired
(5) susceptibility

4. An infant who has never left its mother's side is exposed to people who have colds, but the infant doesn't become ill. Which type of immunity does the infant have?

(1) active immunity—naturally acquired
(2) active immunity—artificially acquired
(3) passive immunity—naturally acquired
(4) passive immunity—artificially acquired
(5) susceptibility

5. A mother wants to avoid the risk of her child's contracting the mumps. The child is given a vaccine that is derived from the mumps virus itself. Which type of immunity does the child now have?

(1) active immunity—naturally acquired
(2) active immunity—artificially acquired
(3) passive immunity—naturally acquired
(4) passive immunity—artificially acquired
(5) susceptibility

6. A patient needs emergency, life-saving treatment after exposure to a deadly virus. Which type of immunity will the patient have?

(1) active immunity—naturally acquired
(2) active immunity—artificially acquired
(3) passive immunity—naturally acquired
(4) passive immunity—artificially acquired
(5) susceptibility

7. A person who contracts AIDS is unable to fight off infection and falls prey to illness after illness. Which type of immunity does such a person have?

(1) active immunity—naturally acquired
(2) active immunity—artificially acquired
(3) passive immunity—naturally acquired
(4) passive immunity—artificially acquired
(5) susceptibility

Item 8 is based on the following equation.

CH_4	+ $2O_2$	→	CO_2	+ $2H_2O$
methane	oxygen		carbon dioxide	water

8. According to this equation, which of the following facts explains why burning a hydrocarbon-fueled lamp in a tent is unsafe, even if combustion is complete?

(1) A gas will be produced.
(2) The fuel may be broken down into elements.
(3) The burning fuel may give off fumes.
(4) Water vapor may build up inside the tent.
(5) The oxygen inside the tent will be consumed.

9. Organisms that contain the chemical chlorophyll, such as green plants, are called *producers*. They are able to make their own food from carbon dioxide and water, using energy from sunlight to fuel the chemical process (photosynthesis). Producers are eaten by consumers, who use the food energy of the producers for their own growth and energy requirements.

The ultimate source of energy for all living things on Earth is

(1) carbon dioxide
(2) chlorophyll
(3) green plants
(4) the sun
(5) water

Items 10 to 12 refer to the following graph.

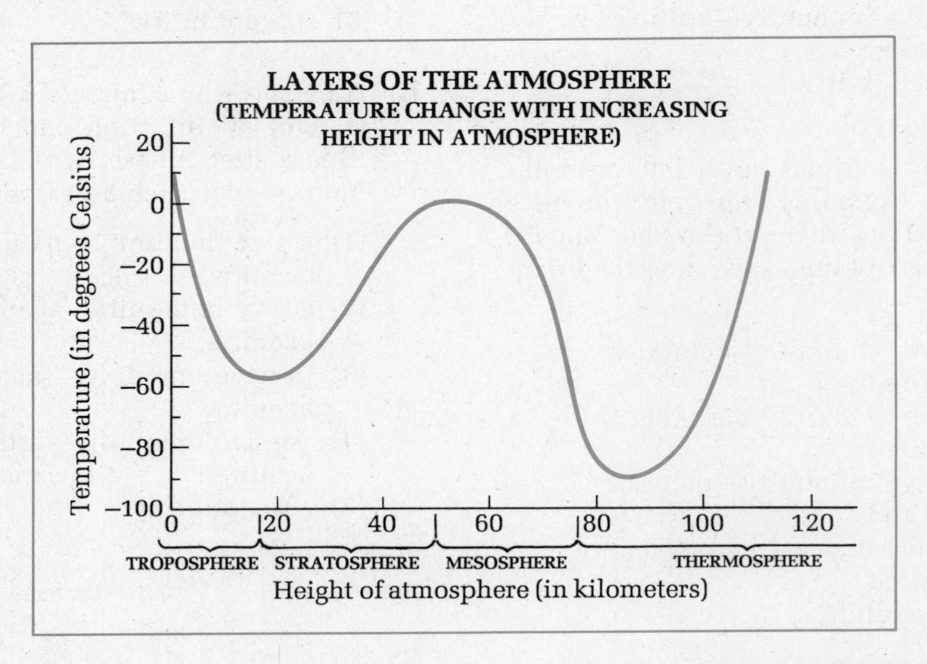

10. In which layer of the atmosphere are the coldest temperatures found?

(1) the surface of Earth
(2) the troposphere
(3) the stratosphere
(4) the mesosphere
(5) the thermosphere

11. The ozone layer of the atmosphere, which is located at the top of the stratosphere, reflects the infrared heat rays of Earth. Certain forms of pollution are increasing the thickness of this layer.

Which of the following phenomena would most likely result from the thickening of the ozone layer?

(1) The temperature of the thermosphere would increase.
(2) The size of the mesosphere would decrease.
(3) The temperature of the troposphere would increase.
(4) The temperature of the stratosphere would decrease.
(5) The thickness of the thermosphere would increase.

12. The temperature of a substance is a measure of the amount of its molecular energy. Why, then, would it be impossible to continue this graph past the thermosphere to determine the temperature in space?

(1) There is no ozone in space.
(2) Space is a vacuum, containing almost no molecules.
(3) It is impossible to determine where the atmosphere ends and space begins.
(4) Space contains molecules that have no energy.
(5) The numbers on the graph cannot go high enough.

---

13. During the development of a human fetus, one cell divides repeatedly to produce millions of specialized cells that take on specific roles in the body. Ectoderm cells form skin and the nervous system, and endoderm cells form deep internal structures such as the digestive system. In between, the mesoderm cells form the skeleton, the muscles, the circulatory system, and other structures.

A person's spinal cord is formed from the

(1) circulatory system
(2) ectoderm
(3) skin
(4) mesoderm
(5) endoderm

Item 14 refers to the following table.

Common Types of Molecules in the Human Body

Molecule	Some Functions
Water	Aids digestion Controls body temperature through perspiration Dissolves and rids body of wastes Transports substances in blood
Proteins	Increase the speed of chemical reactions Build cells and body structures
Fats	Store energy for long-term use Protect organs
Carbohydrates	Provide energy for immediate use Provide the chief source of energy for muscle contraction
Minerals	Allow proper functioning of certain proteins Maintain water balance

14. A long-distance runner finishes a 20-mile race on a hot day in mid-July. According to the table above, which of the following types of molecules would you expect to be most depleted as a result of this exertion?

(1) carbohydrates and fats
(2) carbohydrates and water
(3) fats and proteins
(4) minerals and proteins
(5) proteins and water

Items 15 to 18 are based on the following passage.

Motion can be thought of as the process of changing position; an object that is moving is going from one location to another or, perhaps, is tilting to face a different direction. Motion is a familiar part of our everyday lives: we see the sun rise and set as Earth moves; we use a car or a bus to move us to work; we use a fork to move food to our mouths.

The motion of any object has both a speed and a direction. *Speed* can be described as the distance that an object moves in a certain period of time. Speed may vary from moment to moment, but the average speed of a moving object can be determined by dividing the total distance that is traveled by the amount of time that the trip took. A moving object will move in a straight line at a constant speed unless an outside force such as friction causes it to change speed or direction. Any change in speed or direction is called *acceleration*; slowing down, speeding up, or turning are forms of acceleration.

15. A truck driver traveled 1800 miles in 30 hours. Which of the following statements concerning the driver's speed could NOT be true?

(1) The driver's speed often fell below 20 mph.
(2) The driver's average speed was 40 mph for the first 10 hours of the trip.
(3) The driver did not exceed the interstate speed limit of 55 mph.
(4) The driver's speed was below 60 mph for more than half of the trip.
(5) The driver traveled at 70 mph for most of the trip.

16. According to the passage above, Earth is accelerating as it orbits the sun because

(1) its speed is increasing
(2) it is not moving in a straight line
(3) it is slowing down
(4) the distance covered is not constant
(5) the time it takes to orbit the sun is not constant

17. If a rubber ball is thrown from a space shuttle into space, it will travel in a straight line and at a constant speed. What is the best explanation for this phenomenon?

(1) Acceleration is not possible in space.
(2) The ball's average speed will constantly increase.
(3) There is no outside force to slow or to stop the ball.
(4) There is no such thing as "direction" in space.
(5) Moving objects in space have no speed.

18. From the given facts, you can conclude that a passenger in a car crash is flung forward because

(1) the speed of moving objects can vary from moment to moment
(2) any object is in motion when it changes position
(3) an object can change both its speed and its direction at the same time
(4) moving objects continue to move unless a force stops them
(5) an increase in speed is one form of acceleration

---

19. *Rusting* is the common term to describe the chemical reaction that occurs when iron is combined with oxygen. Rust—a reddish-brown chemical, hydrated iron oxide—is produced more quickly in the presence of moisture.

A museum has acquired a rare iron tool that is thousands of years old. Which storage environment will most minimize the possibility of further rusting?

(1) dry and in the open air
(2) dry and sealed in a vacuum
(3) damp and in an airtight container
(4) damp and in a dark room
(5) damp and in the sun

20. Many of our physical characteristics are determined by our genes, which we inherit from our parents. The genes for each trait come in pairs, one from each parent. Often, one form of a gene can cause a disease, but only if a child receives that form from both parents. Sickle-cell anemia is an example of such a disease: a child will have the anemia only if he or she inherits the disease-causing gene from both parents.

Two potential parents, both of whom suffer from sickle-cell anemia, visit a genetic counselor. They are told that their chances of having children who are free from the disease are

(1) 100%
(2) 75%
(3) 50%
(4) 25%
(5) 0%

Items 21 to 25 are based on the following passage.

The function of the digestive system is to break food down into molecules that are small enough to be absorbed through the wall of the digestive tract into the bloodstream. The blood takes these small molecules to the other parts of the body, where they are used as suppliers of energy to power metabolic reactions or as building blocks to replace worn-out cells or structures. Breaking down food into small molecules involves two processes: mechanical digestion, such as chewing and grinding, and chemical digestion, in which chemicals are used to break down large food molecules.

The food we eat passes through a series of connected organs that form a tube about 30 feet long. Digestion begins in the mouth with chewing and with the chemical action of saliva on the food. When swallowed, the partially digested food moves down the 10-inch tube called the *esophagus* to the stomach, where it is further broken down mechanically and chemically.

From the stomach, the food enters the small intestine, where chemical digestion is completed. The food has now been broken down into small molecules that can enter the bloodstream through the wall of the small intestine and be carried to other organs in the body.

Undigested food enters the large intestine, where water is removed and returned to the blood. Bacteria that live in the large intestine produce vitamins that are absorbed along with the water. The remaining undigested material is excreted through the anus.

21. In which organ of the digestive system does chemical breakdown begin?

(1) the mouth
(2) the esophagus
(3) the stomach
(4) the small intestine
(5) the large intestine

22. A person who is suffering from the flu has diarrhea. Which organ is most likely to be NOT working properly?

(1) the esophagus
(2) the stomach
(3) the small intestine
(4) the large intestine
(5) the anus

23. The inside walls of the small intestine are not smooth. They are lined with fingerlike projections called *villi*, which increase the surface area of the intestinal walls. Of what benefit is the increased surface area?

(1) It allows more water to be absorbed into the blood.
(2) It provides more surface area for the mechanical breakdown of food.
(3) It increases the area for the absorption of molecules into the blood.
(4) It allows food to pass through the small intestine more quickly.
(5) It allows the digestive system to process larger pieces of food.

24. In which of the following organs does no breakdown of food take place?

(1) mouth and esophagus
(2) stomach and anus
(3) mouth and small intestine
(4) esophagus and large intestine
(5) small intestine and large intestine

25. Antibiotics kill bacteria. The passage suggests that taking antibiotics could cause which of the following side effects?

(1) malnutrition
(2) allergic reaction
(3) drowsiness
(4) nausea
(5) loss of appetite

Item 26 is based on the following figure.

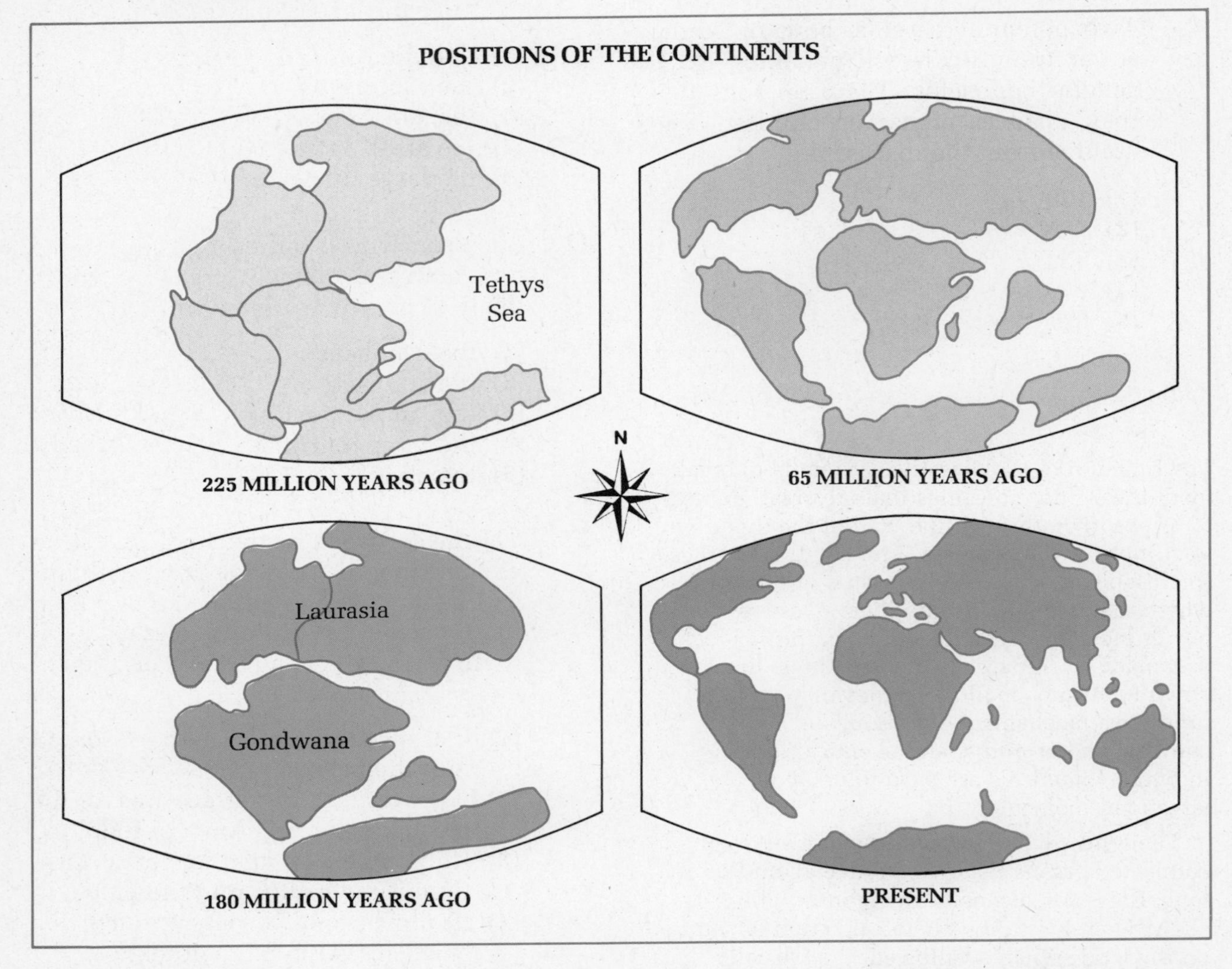

26. The southern continents show some similarities between their plant and their animal communities. The fossil record shows that the communities were even more alike millions of years ago. Which of the following conclusions about these phenomena is best supported by information in the figure above?

(1) Evolution occurs slowly in the Southern Hemisphere.
(2) The southern continents once were joined, and organisms moved freely across them.
(3) The huge Tethys Sea kept all the organisms close to each other.
(4) Many of the organisms ended up in Laurasia, not Gondwana.
(5) A few million years ago, evolution had just begun and there were not many kinds of organisms.

27. *Cold* is really the absence of heat energy in a substance. A can of soda that is put into a refrigerator does not gain coolness; it loses heat, which is absorbed by its surroundings. Energy always flows from substances that are high in energy to substances that are lower in energy.

When an iron pan that has been heating on the stove is filled with cold water, which of the following occurs?

(1) Cold moves from the water into the pan.
(2) Heat flows from the pan to the water.
(3) The water loses heat to the pan.
(4) Energy moves from the water to the pan.
(5) No energy is transferred at all.

28. Air pollutants that are carried by winds from industrial areas undergo chemical reactions in the atmosphere that result in the formation of acid rain. The increased acidity of the rain damages the leaves and the stems of plants and may interfere with plant reproduction. These problems presently threaten many major forests of the northeastern United States.

Which of the following actions could be the most effective way of decreasing the amount of acid rain in the northeastern states?

(1) Spray forests with chemicals that neutralize the acid.
(2) Clean up the circulating air at its northern sources.
(3) Give the forest plants time to adapt to the increased acid levels.
(4) Allow the forests to die, and replant with resistant species.
(5) Treat damaged plants immediately after exposure.

Items 29 and 30 are based on the following passage.

An electrical circuit is characterized by the flow of electrons. In a series circuit, the electrons leave the negative pole and flow, one by one, through each electrical device before they reach the positive pole of the source. In a parallel circuit, the electrons may flow side by side. As a result, the current can be divided; the electrons can flow through different paths to each electrical device.

All electrical circuits are either series circuits or parallel circuits. The major difference between the two is what happens to the current if a device in the circuit is disconnected. If an electrical device in a parallel circuit is disconnected, all other electrical devices on the same circuit will continue running because each has its own pathway for the current flow. If an electrical device in a series circuit is disconnected, all other devices will stop running because the circuit is broken.

29. A string of lights on a Christmas tree goes out, but all the other strings not in the same plug remain lit. Which of the following best explains what has happened?

(1) The string is wired in parallel, and one light bulb has burned out.
(2) The string is wired in series, and one light bulb has burned out.
(3) All the light bulbs on the string have burned out at one time.
(4) The poles have changed from negative to positive.
(5) The current has changed from AC to DC.

30. In which direction is the flow of electrons in an electrical current?

(1) from the negative terminal to the positive terminal
(2) from the positive terminal to the negative terminal
(3) back and forth between the positive terminal and the negative terminal
(4) from positive to negative in a series circuit; from negative to positive in a parallel circuit
(5) from the negative terminal in one device to the negative terminal in the next device

Items 31 to 33 are based on the following graph.

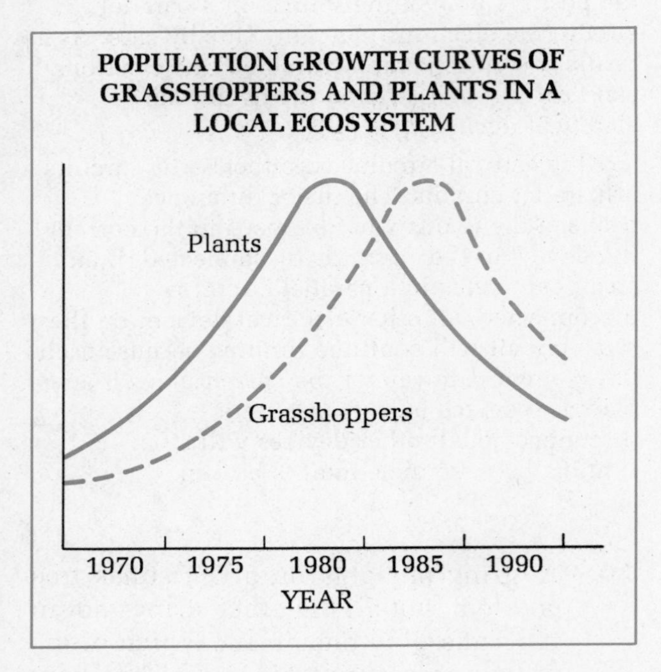

31. According to the graph above, what will happen to the grasshopper population in 1995?

(1) All the grasshoppers will die.
(2) The population will increase slowly.
(3) The population size will increase slowly in one year.
(4) The population will continue to decrease.
(5) The population will level off.

32. Which of the following conclusions can be supported by information in the graph?

(1) The grasshopper population constantly seeks to rise above the ability of the environment to sustain it.
(2) Grasshoppers were introduced into the ecosystem sometime around 1965, when the ratio of plants to grasshoppers would have been 1:1.
(3) Because the plant life and the animal life in this ecosystem parallel each other, the ecosystem is doomed to extinction.
(4) The amount of plant life in an ecosystem is the sole determinant for the amount of animal life.
(5) The grasshopper and the plant populations influence each other's tendency to increase or to decrease.

33. A household has problems with mice. Which of the following suggestions would likely prove the most effective in decreasing the mouse population over a long period of time?

(1) Eliminate the younger mice first.
(2) Seal up food so that the mice cannot get at it.
(3) Bring in at least one cat to prey upon the mice.
(4) Set out mouse traps.
(5) Set out poison.

---

34. When crystalline gypsum is heated, water vapor is driven off and a white powder, plaster of paris, is produced. When water is added to plaster of paris, a hard, white solid (which is used to make casts for broken bones) can be formed.

According to this passage, how can you tell that the chemical reaction that results when gypsum is heated is irreversible?

(1) Heat is not given off when water is added to plaster of paris.
(2) It is impossible to add water vapor to plaster of paris.
(3) A crystalline substance is not formed when water is added to plaster of paris.
(4) You do not need to add energy to produce the hard, white solid.
(5) Plaster of paris hardens very quickly, regardless of the application of heat.

Items 35 to 38 are based on the following passage.

Although life has been present on Earth for more than 3 billion years, detailed records of organisms on Earth began only about 600 million years ago. The period of time before that is called the *Precambrian Era*. Organisms then were simple and soft bodied.

The era after the Precambrian is called the *Paleozoic*, and it lasted for about 350 million years. During this period of time, most forms of life appeared. The Paleozoic thus is often divided into smaller periods, each named for the animal group that was dominant at that time. The Age of Invertebrates was followed by the Age of Fishes (the first animals with backbones), which was succeeded by the Age of Amphibians (the first land-dwelling animals). During this time, much of modern North America and Europe were underwater, and the Appalachian Mountains were formed. In swampy regions, plants decayed in what in millions of years would become coal deposits.

The *Mesozoic Era* (the Age of Reptiles) followed the Paleozoic, and it lasted for about 150 million years. Dinosaurs thrived, and the first small mammals and flowering plants appeared. The Rocky Mountains started to become stable. Dinosaurs disappeared at the end of this era.

The *Cenozoic Era*, often called the Age of Mammals, began 60 million years ago. The oldest ancestors of modern primates appeared 1 to 3 million years ago in Africa. This era has been marked by the spread and retreat of glaciers during the Ice Age.

35. During which period of time did the first vertebrates appear?

(1) Precambrian Era
(2) Paleozoic Era
(3) Age of Amphibians
(4) Age of Reptiles
(5) Cenozoic Era

36. Many fossils are displayed at a certain museum. If the displays reflect the actual availability of fossils in the world, which period of time will be the least represented?

(1) Precambrian Era
(2) Age of Invertebrates
(3) Age of Amphibians
(4) Mesozoic Era
(5) Cenozoic Era

37. Fossils of an organism that lived in the ocean are found in exposed rocks of the eastern coastal plain of North America. During which period of time were these fossils probably formed?

(1) Precambrian
(2) Paleozoic
(3) Age of Reptiles
(4) Cenozoic
(5) the present

38. Often, predictions can be made on the basis of trends. Which of the following predictions is consistent with trends that are seen in the fossil record that is described above?

(1) Eventually, conditions will return to those of the Precambrian Era.
(2) Invertebrates will become the dominant animals.
(3) Living creatures that are unknown today will walk on Earth.
(4) New species will develop only in the event of another ice age.
(5) Fewer and fewer fossils will be formed in future years.

Item 39 is based on the following drawing.

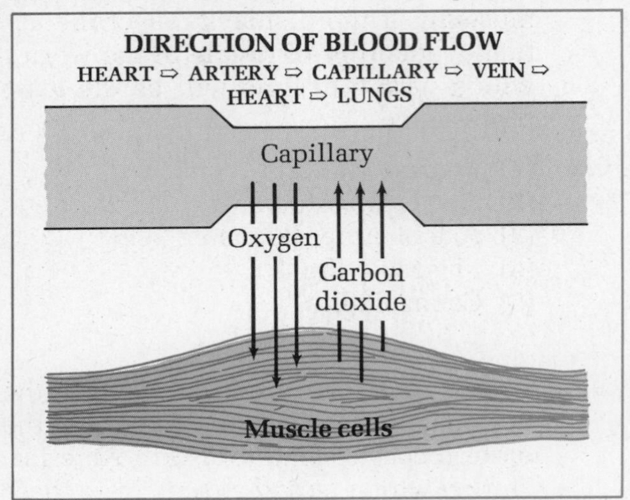

39. The figure above illustrates the exchange of oxygen and carbon dioxide between the blood and the muscle cells. Blood always has the highest ratio of oxygen to carbon dioxide when it is traveling

(1) from the heart to the arteries
(2) from the muscle cells to the veins
(3) through the veins to the heart
(4) in the muscle cells
(5) from the veins to the lungs

40. Earthworms are *hermaphroditic*: each individual earthworm has both male and female sex organs and produces both eggs and sperm. However, an earthworm never fertilizes its eggs with its own sperm; in fact, it is unable to do so. Two worms come together and exchange sperm, which then fertilize the eggs.

Which of the following statements explains why it is more advantageous to the species for earthworms to exchange sperm instead of fertilizing their own eggs?

(1) Self-fertilization is physically impossible.
(2) A worm produces many more sperm than eggs.
(3) Exchanging sperm provides better protection for the two worms during mating.
(4) Sperm are unable to fertilize eggs that are produced by the same individual.
(5) The exchange of sperm allows for the mixing of genes, which results in healthier offspring.

Items 41 to 44 are based on the following table.

Human ABO Blood Types

Blood Type	Genetic Makeup	Can Donate Safely to:	Can Receive Safely From:
A	AA or AO	A or AB	A or O
B	BB or BO	B or AB	B or O
O	OO	O, A, B, or AB	O
AB	AB	AB	O, A, B, or AB

41. Which of the following blood types could be given safely to a type B person?

(1) B only
(2) B or AB
(3) B or O
(4) O only
(5) all types

42. If you were to set up a blood bank to supply blood to persons in need, which type(s) of blood would be the most useful to keep in supply?

(1) A
(2) B
(3) AB
(4) O
(5) A and B

43. If a child's mother is blood type A and the child's father is type O, which blood type(s) might the child have?

(1) A only
(2) A or O
(3) A or AB
(4) AB or O
(5) AB only

44. It is not difficult for type O persons to find blood when they need transfusions. Which of the following explanations supports the preceding statement?

(1) Most people are type O, so supplies are plentiful.
(2) Most people who are type A or type B have AO or BO genotypes.
(3) Type O can receive blood from any other type.
(4) It is not dangerous for a type O person to receive blood of another type.
(5) Type AB parents always have type O children.

---

45. When a population is too large to study, information about it can be gathered by *sampling*. In this technique, small portions of the population are studied, and conclusions that are based on them are assumed to hold for the entire population. Sampling is often used to study and to predict the behavior of natural populations, as in predictions of how voters will act on election day. However, if the conclusions are to be valid for an entire population, which of the following conditions must be true?

(1) The sample must include more than half the population.
(2) The sample must be representative of the population.
(3) Each member of the sample must be average in all respects.
(4) The study must not take a long period of time.
(5) All members of the sample population must live near each other.

Items 46 and 47 are based on the following passage.

Moist air over the Pacific Ocean is carried eastward across the United States by the prevailing winds. As the air rises up and moves over the western mountains, it cools and drops much of its moisture. By the time the air has crossed the peaks, it has little moisture left and creates a "rainshadow" on the eastern slope.

46. According to this paragraph, which of the following regions would be the most promising for farming?

(1) the shoreline of the ocean
(2) valleys near the ocean
(3) the foothills that rise eastward into the mountains
(4) the eastern slope of the mountains
(5) the plains east of the mountains

47. Airplanes that fly with a tail wind move more quickly than those with a head wind. Which of the following airplane trips would take the longest?

(1) New York to San Francisco
(2) San Francisco to New York
(3) Portland to St. Louis
(4) Seattle to Chicago
(5) Chicago to Seattle

Item 48 refers to the following graph.

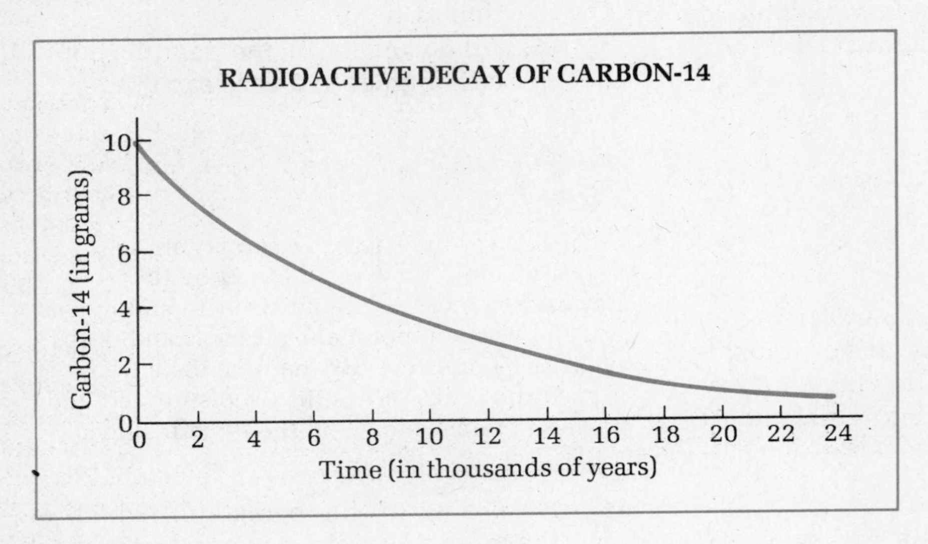

48. The half-life of a radioactive substance is the amount of time it takes for half of a sample to decay. What is the approximate half-life of carbon-14?

(1) 5 years
(2) 200 years
(3) 2,000 years
(4) 6,000 years
(5) 18,000 years

Items 49 to 52 are based on the following passage.

One of the basic principles of biology is that all living things are made of cells, the smallest units of life. The simplest organisms, such as bacteria or miscroscopic pond creatures, consist of only one cell. More complex organisms are made up of millions or even trillions of cells of various shapes and sizes that reflect their functions. Often, these cells are grouped together into tissues and organs that have specific functions in the body.

Red blood cells are round, flattened discs. This shape allows an iron-containing chemical in the cell to pick up and deliver oxygen to other cells of the body quickly. White blood cells can change their shape. This ability enables them to squeeze between other cells to attack infecting organisms and other foreign particles. Nerve cells are generally very long and thin. They carry to and from the brain signals that coordinate the body's activities. Muscle cells are elastic and can shorten to pull on bones and to cause such motion as walking or writing.

A single-celled organism can perform many of the same functions as more complex creatures can. Organelles, structures within cells, have roles such as breaking down food or getting rid of wastes. The cell membrane allows the organism to absorb oxygen, to move, and to respond to changes in its environment.

49. Which cells cause the muscle cells to shorten?

(1) bacteria
(2) red blood cells
(3) nerve cells
(4) other muscle cells
(5) bone cells

50. A person who suffers from anemia has "iron-poor blood." In this condition, the cells of the body are probably NOT

(1) being protected well enough by the white blood cells
(2) receiving messages quickly enough from the brain
(3) carrying out their normal function
(4) receiving enough oxygen from the red blood cells
(5) maintaining their proper shapes

51. A virus consists of a molecule of nucleic acid that is surrounded by a coat of protein. Even though it reproduces, a virus generally is not considered to be a living organism.

Which of the following statements best supports this classification?

(1) A virus is not a cell, nor is it made up of cells.
(2) A virus can reproduce only within another organism.
(3) A virus cannot change its shape.
(4) A virus has no nervous system.
(5) Bacteria are the smallest living things.

52. In which of the following ways are organelles similar to organs?

(1) Both are very small.
(2) Both have specialized functions in an organism.
(3) Both are made up of many cells.
(4) Both are building blocks for other structures in the body.
(5) Both can be found in single-celled animals.

---

53. Plants that are pollinated by bees share many characteristics. Most of them have bright colors and sweet odors, both of which characteristics attract bees. Also, they are open during the day, when bees are active.

Bees visit these flowers as they gather nutritious nectar to produce honey. As the bees collect the nectar, pollen clings to their feet. As the bees move from flower to flower, they "track" the pollen, leaving the pollen of one flower on another so that the plants can reproduce.

Which of the following statements is NOT supported by this passage?

(1) Dull, odorless plants are not likely to attract bees.
(2) While fullfilling their own needs, bees perform a service to plants.
(3) Many brightly colored flowers that have sweet odors are pollinated by bees.
(4) Bees do not pollinate plants at night.
(5) The evolutions of both bees and plants have been closely connected.

Items 54 and 55 are based on the following passage.

Lines of longitude run from north to south and measure distance in an east-west direction. All extend from the North Pole to the South Pole and cover 180 degrees of the globe.

Lines of latitude run from east to west around the globe and measure distance north and south of the equator. They vary in length (being smaller toward the poles), but they are parallel.

54. Map projections try to represent the round Earth on flat paper. Greenland, near the North Pole, generally appears to be much larger on maps than it really is. Based on the information given, what is the best explanation for this phenomenon?

(1) When the globe is "flattened," the lines of longitude are spread apart at the poles.
(2) Map projections require more than 180 degrees.
(3) Lines of latitude can no longer be parallel on a flat surface.
(4) Greenland has never been correctly mapped.
(5) The North and South poles are closer together on a flat surface.

55. Which of the following tasks could be performed by a crew of sailors who use only latitude and longitude lines?

(1) measuring the distance between ports
(2) identifying stars in the sky
(3) stating their location on the sea
(4) estimating the travel time between ports
(5) measuring the length of the equator

Items 56 to 59 are based on the following passage.

The pH scale measures the acidity of a substance on a scale from 0 to 14. A pH of 7 is neutral; a substance that has this pH is neither an acid nor a base. A pH of less than 7 is acidic; lower numbers will indicate stronger acids. Numbers that are higher than 7 indicate stronger bases. Each number on the scale represents a tenfold difference in acidity; thus, pH 5 is 10 times more acidic than pH 6, and pH 11 is 100 times more basic (or less acidic) than pH 9.

The pH scale is often useful to indicate the amount of acidity in natural substances. For example, rainwater that tests at a pH of 6 is normal and safe; acid rain, however, can be 100 times stronger. Although our stomachs are normally acidic (pH 3 to pH 2), some foods may lower the pH further: sodium bicarbonate is often prescribed to neutralize an overly acidic stomach. Soil on a forest floor often tests at a low pH because decaying evergreen needles release acids into the soil; many plants cannot be grown in this acidic soil, but others, such as cranberries or blueberries, thrive in it.

56. If sodium bicarbonate neutralizes acids, it is probably

(1) a stronger acid
(2) water
(3) a substance that has a pH of 7
(4) neither an acid nor a base
(5) a base

57. The pH of acid rain is

(1) 2
(2) 4
(3) 5
(4) 6
(5) 8

58. When an acid and a base of equal strength are combined, a chemical reaction occurs. What would you expect the pH of the resulting solution to be?

(1) 2
(2) 4
(3) 7
(4) 10
(5) 12

59. The blueberries in an orchard plot are now growing well. According to the passage, which of the following steps might help them to grow?

(1) Water the bushes more often.
(2) Fertilize the bushes with sodium bicarbonate.
(3) Mulch the soil with pine needles.
(4) Plow the soil.
(5) Raise the pH of the soil.

---

60. Blood is a tissue that consists of fluid plasma and the loose blood cells that are suspended in this plasma. Blood types A, B, and AB contain proteins in their red blood cells that cause the plasma of other blood types to form antibodies and then to clot. If type A whole blood is mixed with type B, the type B will clot in response to the type A proteins. Blood type O has no proteins, so it will form antibodies to all other types.

A victim of a motorcycle accident needs blood immediately, but there is none of the patient's type, O. What would be the safest action for the doctor to take?

(1) Give the patient a transfusion of either blood type A or B.
(2) Treat the patient with proteins.
(3) Inject the patient with antibodies.
(4) Give the patient a transfusion of plasma alone.
(5) Wait until type O blood is available.

Items 61 to 63 are based on the following chart.

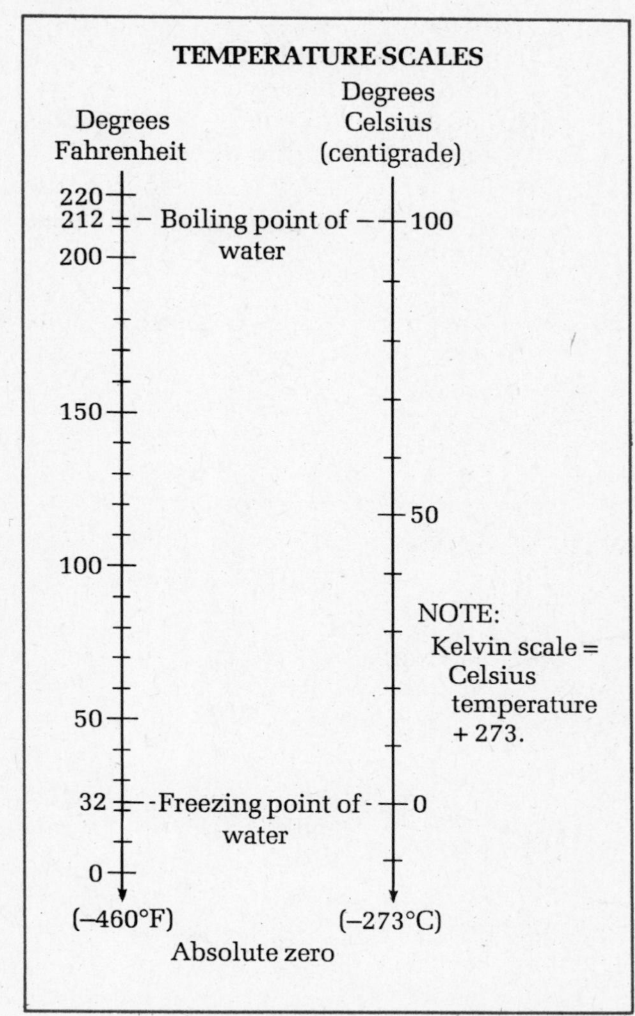

61. At 0 degrees Farenheit, a Celsius thermometer would read closest to which of the following?

(1) 10 degrees
(2) 0 degrees
(3) −20 degrees
(4) −32 degrees
(5) absolute zero

62. Which of the following statements is true of a comparison between the Celsius and the Kelvin temperature scales?

(1) Both scales use the same number to designate absolute zero.
(2) Both scales measure temperatures above and below a fixed point of 0 degrees.
(3) The degrees on both scales are measured at the same interval.
(4) Neither scale uses numbers as great as those on the Farenheit scale to measure temperature.
(5) The Celsius scale provides more accurate measurements of temperature than does the Kelvin scale.

63. What is the boiling point of water on the Kelvin scale?

(1) −40°
(2) 100°
(3) 212°
(4) 373°
(5) 672°

Items 64 and 65 are based on the following passage.

Both loudness and pitch are determined by characteristics of a wave of sound. Loudness is due to the amplitude of a wave—the distance of the trough or crest of a wave from its midline. The greater the distance, the louder the sound will be. As waves move away from the source of a sound, the amplitude decreases.

Pitch is due to the frequency of vibration—the number of waves that are completed in one second. The higher the frequency, the higher the pitch will be.

64. Imagine that you are walking in your neighborhood and you hear a very loud radio a block away. You are a fair distance from the radio. What accounts for the volume of the sound?

(1) The frequency of the sound has remained undisturbed.
(2) The frequency of the sound has risen.
(3) Air molecules have slowed the dissipation of the waves of sound through the air over a short distance.
(4) The amplitude of the waves of sound is decreasing.
(5) The source has enough energy to sustain the amplitude over a greater distance.

65. In an experiment, a C tuning fork is struck, and the vibration of its prongs is measured. The prongs are found to vibrate 256 times each second. The tuning fork is struck several times; the measurement is the same each time. Which of the following principles that concern sound has been illustrated in this experiment?

(1) Pitch changes in direct relationship to the amount of energy in the source.
(2) Loudness changes in direct relationship to the amount of energy in the source.
(3) Energy is absorbed by a source to produce a softer sound.
(4) The amount of energy in the source does not affect the pitch.
(5) The more energy in the source, the greater the frequency of the waves will be.

Item 66 is based on the following figure.

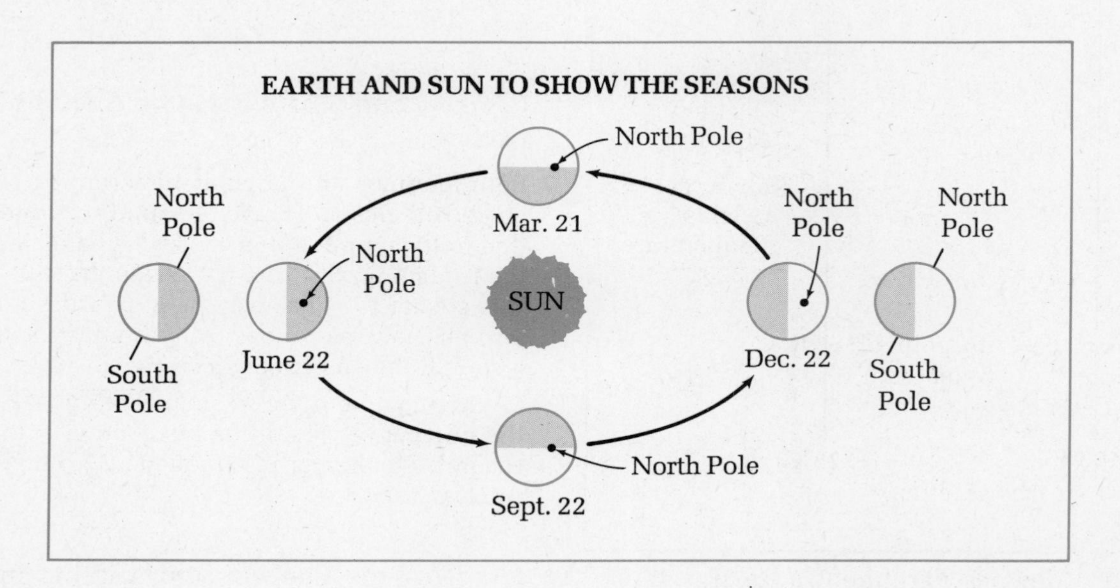

66. Based on the figure, which of the following phenomena should occur in the Southern Hemisphere as the North Pole tilts toward the sun?

(1) The sun will not drop below the horizon.
(2) The growing season will lengthen.
(3) The hours of daylight will decrease.
(4) The hours of daylight will slowly increase.
(5) The pull of the tides will grow stronger.

Answers are on pages 273–275.

PRACTICE TEST

Performance Analysis Chart

Directions: Circle the number of each item that you got correct on the Practice Test. Count how many items you got correct in each row; count how many items you got correct in each column. Write the number correct per row and column as the numerator in the fraction in the appropriate "Total Correct" box. (The denominators represent the total number of items in the row or column.) Write the grand total correct over the denominator, **66** at the lower right corner of the chart. (For example, if you got 59 items correct, write 59 so that the fraction reads 59/**66.**) Item numbers in color represent items based on graphic material.

Item Type	*Biology (page 83)*	*Earth Science (page 106)*	*Chemistry (page 123)*	*Physics (page 142)*	TOTAL CORRECT
Comprehension (page 31)	9, 13, 21, 24, 35, 37, 41, 52	10	56	16, 30, 48, 49, 61	**/15**
Application (page 39)	3, 4, 5, 6, 7, 14, 20, 22, 33, 42	46, 47, 55	19, 57	1, 15 27, 29	**/19**
Analysis (page 46)	23, 31, 36, 39, 43, 50, 60	2, 11, 54, 66	8, 34, 58, 59	17, 62, 63, 64	**/19**
Evaluation (page 57)	25, 28, 32, 38, 40, 44, 45, 51, 53	12, 26		18, 65	**/13**
TOTAL CORRECT	**/34**	**/10**	**/7**	**/15**	**/66**

The page numbers in parentheses indicate where in this book you can find the beginning of specific instruction about the various fields of science and about the types of questions you encountered on the Practice Test.

Simulation

Introduction

The preceding sections of this book have one purpose—to prepare you for the Science Test. By this time, you are probably asking yourself, "Am I ready to take the GED Test?" Your score on the following Simulated Test will help you answer that question.

The test is as much like the real Science Test as possible. The number of questions and their degree of difficulty are the same as on the real test. The time limit and the mixed order of test items are also the same. By taking the Simulated Test, you will gain valuable test-taking experience and get a better idea about just how ready you are to take the actual test.

Using the Simulated Test to Your Best Advantage

There is only one way you should take the simulated test, unlike the practice activities. You should take the test under the same conditions you will have when you take the real test.

- When you take the GED, you will have 95 minutes to complete the test. Though this will probably be more than enough time, set aside at least 95 minutes so you can work without interruption.
- Do not talk to anyone or consult any books as you take the test. If you have a question about the test, ask your instructor.
- If you are not sure of an answer, take your best guess. On the real GED, you are not penalized for wrong answers. Guessing a correct answer will better your score, whereas guessing a wrong answer will not affect your score any more than not answering.

As you take the Simulated Test, write your answers neatly on a sheet of paper or use an answer sheet provided by your teacher. When time is up, you may wish to circle the item you answered last, then continue with the test. This way, when you score your test, you can see how important time was in your performance.

Using the Answer Key

Use the answers and explanations (page 276) to check your answers. marking each item you answered correctly. Regardless of whether you answer an item correctly, you should read the information that explains each correct answer. This will reinforce your test-taking skills and your understanding of the material.

How to Use Your Score

If you got 53 items or more correct, you did 80 percent work or better. This shows that you are most likely working at a level that would allow you to do well on the actual Science Test. If you got a slightly fewer than 53 items correct, you probably need to do some light reviewing. If your score was far below the 80 percent mark, you should spend additional time reviewing the lessons that will strengthen the areas in which you were weak. The Skills Chart at the end of the test will help you identify your stronger and weaker areas.

SIMULATED TEST

TIME: *95 minutes*
Directions: *Choose the one best answer to each question.*

Item 1 refers to the following figure.

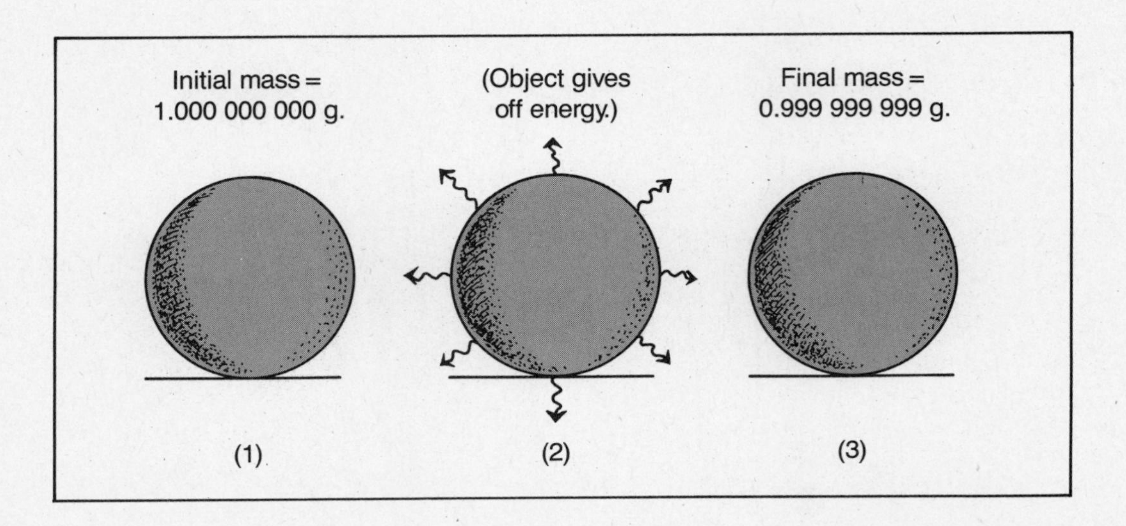

1. Which of the following statements best summarizes the information in the diagram?

(1) Only certain objects can convert their energy into mass.
(2) The volume of an object is noticeably smaller after the object gives off energy.
(3) A considerable amount of mass is lost by an object when the object gives off energy.
(4) A very small amount of mass is lost by an object when the object gives off energy.
(5) Measurements of objects that have a mass of 1.0 gram or less are unreliable.

2. Pure, distilled water has a density of 62.4 pounds per cubic foot. Seawater, however, is not as pure as distilled water. Seawater has many different salts, which are denser than water, dissolved in it. These salts and other dissolved substances are being carried continually from the land into the sea by rivers and runoff water.

Which of the following conclusions about seawater is implied by this information?

(1) The density of seawater is 62.4 pounds per cubic foot.
(2) As seawater evaporates, the water left behind becomes less dense.
(3) Seawater is less dense than pure water.
(4) River water is denser than seawater.
(5) Seawater is denser than pure water.

Items 3 to 7 refer to the following passage.

An organism can be studied at a number of different levels of organization. This organizational hierarchy provides investigators with a framework in which to view the structure of living things. Listed below are these levels of organization.

(1) compound: Matter consists of atoms that are chemically bonded to form compounds. One of the most important compounds for life is water, H_2O, but other important compounds are proteins, fats, sugars, and nucleic acids.

(2) cell: The cell is the simplest unit that displays all the properties of life. The properties of life are the ability to use energy, to grow, to respond to changes in the environment, to reproduce, and to move in some way. All living organisms are composed of cells, although some of the simplest creatures, such as bacteria, consist of only one cell.

(3) tissue: In multicellular organisms, similar cells work together as a unit that is known as a **tissue,** to perform a specific function. Muscle cells, for example, work together to move a limb.

(4) organ: Two or more tissues combine to form an organ, which carries out a specific activity. The heart, for example, includes muscle tissue that contracts, nerve tissue that triggers the contraction, and epithelial tissue that protects the heart.

(5) organ system: Several organs that work to perform a function in common constitute an organ system. The heart works with arteries and veins, for instance, to transport nutrients, gases, and wastes throughout the body.

Each of the following items describes a situation that refers to one of the five categories defined above. For each item, choose the one category that best describes the situation. Each category may be used more than once in the set of items, but no question has more than one best answer.

3. The roots of a carrot contain a group of cells called *parenchyma*, whose function is to store food and water.

Parenchyma would be classified as a(n)

(1) compound
(2) cell
(3) tissue
(4) organ
(5) organ system

4. Skin consists of several cell layers, each of which contains specialized cells. Some of these cells produce hair and nails, others sense pain, and still others produce sweat.

Skin would be classified as a(n)

(1) compound
(2) cell
(3) tissue
(4) organ
(5) organ system

5. As a result of a motorcycle accident, the rider suffers damage to the brain, to the spinal cord, and to the nerves in the right arm.

These injuries appear to be most damaging to a(n)

(1) compound
(2) cell
(3) tissue
(4) organ
(5) organ system

6. A virus consists of molecules of nucleic acid that are surrounded by a coat of protein. Most viruses reproduce by taking over a host cell and using it to make new viruses. Outside a host cell, viruses do not exhibit any life processes.

The highest level of organization found in a virus is that of a(n)

(1) compound
(2) cell
(3) tissue
(4) organ
(5) organ system

7. Brown algae live as colonies of similar cells. In these colonies, every cell is capable of exactly the same functions.

Therefore, the highest level of organization found in the brown algae must be that of a(n)

(1) compound
(2) cell
(3) tissue
(4) organ
(5) organ system

Item 8 refers to the following table.

Fermentation	
Requirements	**Products**
Sugar Darkness Enzymes Absence of oxygen Warm temperatures	Alcohol Carbon dioxide Energy

8. A shopper finds that a recently purchased bottle of apple cider has fermented. Which of the following statements best describes the cause-and-effect relationship that may have allowed this change to occur?

(1) The shopper opened the bottle, allowing carbon dioxide to enter and to initiate the production of alcohol.
(2) The cider was left at room temperature for one hour, causing the yeast to produce more fruit sugar.
(3) The shopper failed to store the cider in the dark, so the yeast multiplied rapidly.
(4) The cider was left standing for several hours; as a result, carbon dioxide escaped.
(5) The cider was not stored in the refrigerator, where cool temperatures would have slowed fermentation.

---

9. Many organs show structural features, or adaptations, that help them to perform their functions more effectively. The human brain is greatly enfolded and creased, more so than that of other animals. The small intestine, where nutrients are absorbed into the blood, stretches for 20 feet. Its inside surface is also enfolded. The lungs are made of millions of tiny air sacs, where oxygen is delivered to the blood. The property that all these adaptations have in common is their

(1) small size
(2) large surface area
(3) speed of function
(4) relative weight
(5) great mobility

Items 10 to 12 refer to the following map.

Warm air, which is less dense and holds more moisture, tends to rise over cold air. As the warm air rises, it becomes cooler and holds less moisture.

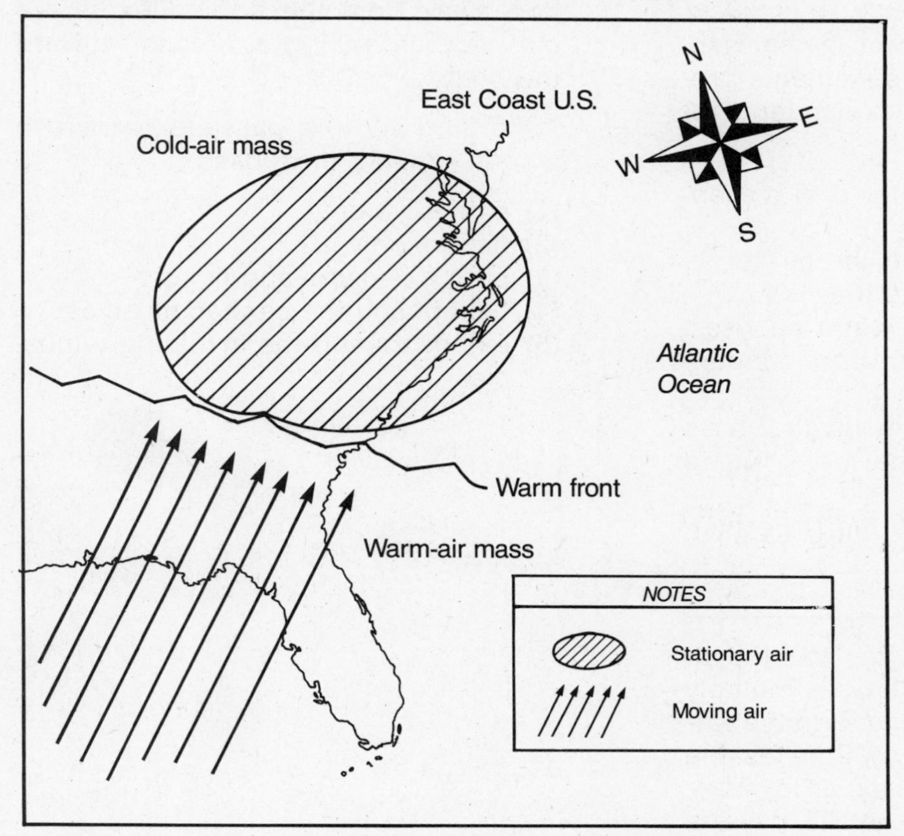

10. If this map represents a typical weather pattern, then a warm front usually develops at a point where

(1) all air, land, and water masses are stationary
(2) a moving warm-air mass meets a stationary warm-air mass
(3) a moving cold-air mass meets a moving warm-air mass
(4) a moving warm-air mass meets a stationary cold-air mass
(5) a moving cold-air mass meets another moving cold-air mass

11. Imagine that this map represents a weather pattern on a Monday night. Which of the following weather conditions will prevail on Tuesday morning in the area where the two fronts meet?

(1) rain
(2) variable cloudiness
(3) temperatures below freezing
(4) no wind
(5) bright sunshine

12. An air-conditioner is run in a room that has a high ceiling. After it has been running for several days, the wallpaper near the ceiling begins to mildew. Based on the information in the map, which of the following statements provides the most logical explanation for the mildew?

(1) Water from a leak in the roof is seeping into the ceiling.
(2) Water from a leak in the air-conditioner is seeping into the wall.
(3) As the room cools, warm air rises, and some of its water vapor condenses on the wall.
(4) Several days of warm weather have caused the air to be more humid than usual.
(5) Cool air holds more moisture than warm air does.

---

13. Human development begins with the division of a fertilized egg. Cells multiply and become specialized to form the various structures of the body. Somites, the beginnings of the muscular system, form in the third week. At 23 days, a weak heartbeat can be detected, and the neural groove has begun to outline the nervous system. By eight weeks, all organ systems are visible, and a human form can be discerned.

According to this passage, when can the digestive system first be seen?

(1) in the third week
(2) at 23 days
(3) as the fertilized egg divides
(4) by eight weeks
(5) at birth

14. Voluntary muscles are those that can be controlled consciously. Extensors are voluntary muscles that pull in a direction away from the body. Flexors are muscles that pull in a direction toward the body.

According to this passage, a person who is eating a meal uses

(1) extensors only
(2) flexors only
(3) both extensors and flexors
(4) flexors and the cardiac muscles
(5) extensors and the involuntary muscles

Items 15 to 17 refer to the following passage.

An object that is thrown horizontally will eventually fall to the earth. As it moves forward, it is pulled to the earth by the force of gravity. The result is a curved trajectory, or path of motion. In the absence of gravity, the thrown object would continue to move horizontally in a straight line. In the absence of friction, its speed would never change. Gravity causes the object to fall, and friction slows it as it moves through the air.

Because it is only the downward pull of gravity that brings the propelled object to the earth, the distance it travels horizontally does not affect the amount of time it takes for the object to hit the ground. For example, any two baseballs that are thrown forward from the same height will hit the ground at the same time, assuming that friction has little or no effect. The acceleration that is caused by gravity is the same for each baseball—or for any other falling object.

An object that is propelled upward and forward, such as a kicked football, also will travel horizontally at a constant speed (again, disregarding friction). Its vertical speed, however, will change during the course of its arch-shaped path. As it moves upward, away from the earth, gravity will slow its ascent. At the peak of its climb, the object has zero vertical velocity. Its horizontal velocity, however, is unchanged: it momentarily is not rising or falling, but it continues forward. The football now will fall to the earth, pulled by gravity. Gravity also will cause it to accelerate as it falls. When caught, its speed, both horizontal and vertical, will be the same as when it was kicked. The time it takes to land again depends only upon how high it is kicked, not upon how far forward it travels.

15. Which of the following is an accurate statement about gravity?

(1) It is a constant for all objects at all times.
(2) It determines how far forward an object travels.
(3) It decreases as an object rises.
(4) It causes a thrown object to slow down.
(5) Its strength depends on air resistance.

16. According to this passage, a thrown object can travel at a constant speed. Which of the following variables does this statement assume?

(1) Gravity has no effect on the motion of the object.
(2) Vertical speed changes; horizontal speed does not.
(3) There is no friction to slow the object.
(4) The object is weightless.
(5) The object is thrown in the absence of gravity.

17. Centerfielder X catches a ball near the wall and throws it in a low arc to second base. Later in the game, Centerfielder Y makes a similar play, but throws the ball horizontally. The ball that is thrown by X reaches its destination, but the ball that is thrown by Y bounces before it reaches second base.

Based on the information in the passage, which of the following is the most scientific explanation for the difference in the outcomes?

(1) Centerfielder X threw the ball more slowly than did Centerfielder Y.
(2) Centerfielder X threw the ball with greater force than did Centerfielder Y.
(3) The density of baseballs inclines them to travel in an arc rather than in a straight line.
(4) The ball that was thrown by Centerfielder X stayed in the air longer than did the ball that was thrown by Centerfielder Y.
(5) The ball that was thrown by Centerfielder X experienced less air friction than did the ball that was thrown by Centerfielder Y.

18. The feeding relationships in a natural community can be viewed as a pyramid, with plants, the producers of food and the most abundant organisms, at the bottom. Plants are eaten by primary consumers (herbivores), who are eaten by secondary consumers (carnivores), who themselves are preyed upon by tertiary consumers (other carnivores). Only about 10% of the food energy that is available at each level of the pyramid can be used by the next; thus, each succeeding level must be much smaller than the one below it. Omnivores, such as humans, draw their food from several layers of the pyramid.

The human population on Earth currently doubles every 40 years. If this trend continues, some resources, such as food, will be in short supply. Based on the information in the passage, which of the following is the best solution to this potential problem?

(1) Increase the consumption of meat.
(2) Use fossil fuels for energy.
(3) Eat more of the primary consumers.
(4) Convert to a vegetarian diet.
(5) Eat less food.

19. In certain areas of the country, the water supply is classified as "hard" or "soft." Hard water contains minerals, such as magnesium and iron, that seep into the groundwater from the soil by a process called leaching. Soapsuds do not form easily in hard water, and the minerals may settle, causing discoloration or buildup. Soft water, on the other hand, contains much smaller amounts of dissolved minerals.

Which of the following statements about hard water is probably based on opinion rather than on fact?

(1) Hard water can cause a ring of discoloration around the inside of a bathtub.
(2) A water softener is needed to wash clothes in hard water.
(3) Soap will lather more easily in soft water than in hard water.
(4) Soft water tastes better than hard water.
(5) Soft water leaves faucets clean and smooth on the inside.

20. Lactose, or milk sugar, is present in most dairy products. Lactase, an enzyme found in the small intestine, is necessary for the digestion of lactose. When people who are "lactose intolerant" drink milk or eat milk products, they often experience diarrhea and painful cramps.

Lactose intolerance occurs when

(1) the body develops an allergy to mother's milk
(2) the body does not produce enough lactase
(3) the enzymes in the intestine interfere with the digestion of dairy products
(4) the lactase present in milk is not digested by bacteria in the intestine
(5) people experience diarrhea and cramps

Items 21 to 25 refer to the following passage.

A vitamin is an organic molecule that is needed by the body in very small amounts. Because the human body cannot make vitamins, vitamins must enter the body through the diet. The amount of a vitamin that is needed on a daily basis is usually measured in milligrams (mg.) The recommended daily dosage of vitamin C, for example, is 200 mg.

Vitamins contribute to the structure of enzymes, the protein molecules that speed up the chemical reactions in the body. Every step of every reaction that occurs in the body—from digestion to protein formation to the release of energy from storage molecules—requires an enzyme. Vitamins are necessary for the proper functioning of some of these enzymes.

Vitamins are classified as either "water soluble" or "fat soluble." Water-soluble vitamins, such as vitamins C and B complex, dissolve in water. Excess water-soluble vitamins are excreted by the body; they cannot be stored. As a result, a body can become deficient in such vitamins quickly. A deficiency in vitamin B_1 causes malfunction of the nerves and weakness of the muscles. A deficiency in vitamin B_2 causes dysfunction of the liver. A lack of vitamin B_6 weakens the immune system; a lack of vitamin C can decrease the body's ability to heal itself when wounded.

Vitamins A, D, E, and K are fat soluble. They can be stored in the body's fat cells. They are not easily excreted; in large doses, they can be harmful. Vitamin A is necessary for night vision; vitamin D strengthens teeth and bones. Vitamin E is important in the formation of blood, and vitamin K helps the blood to clot.

21. A doctor who specializes in sports medicine probably would be most concerned about patients who are deficient in which of the following vitamins?

(1) A
(2) A and B
(3) B_2 and B_6
(4) B_1 and D
(5) C

22. Which of the following procedures should be followed by a person who is preparing to undergo surgery?

(1) Take massive doses of vitamin K.
(2) Discontinue taking vitamin C.
(3) Take moderate doses of vitamins C, E, and K.
(4) Take massive doses of vitamin B_6.
(5) Take moderate doses of vitamins A, C, and D.

23. Manufacturers of vitamins A and D often include cod liver oil in the tablets because

(1) the oil keeps the vitamins from being excreted
(2) the vitamins are best absorbed by the body when dissolved in fat
(3) the oil keeps the body from absorbing too much of the vitamins
(4) the presence of fat speeds up the enzyme process
(5) most people who take vitamins A and D are deficient in cod liver oil

24. Severe, or "crash" diets do not provide enough essential vitamins. Which of the following is the most likely result of such a diet?

(1) Chemical reactions in the body speed up uncontrollably.
(2) Certain bodily functions may slow down or stop entirely.
(3) The body will spontaneously produce enzymes.
(4) The body will spontaneously produce the vitamins that are deficient.
(5) The person will urinate excessively.

25. Which of the following statements is NOT supported by the passage?

(1) It is possible to become poisoned by vitamins.
(2) Taking massive doses of vitamin C is useless.
(3) Deficiencies of fat-soluble vitamins are more common than deficiencies of water-soluble vitamins.
(4) It is possible to determine whether one is receiving enough of certain vitamins.
(5) A person who lacks a particular enzyme may have a vitamin deficiency.

Item 26 refers to the following map.

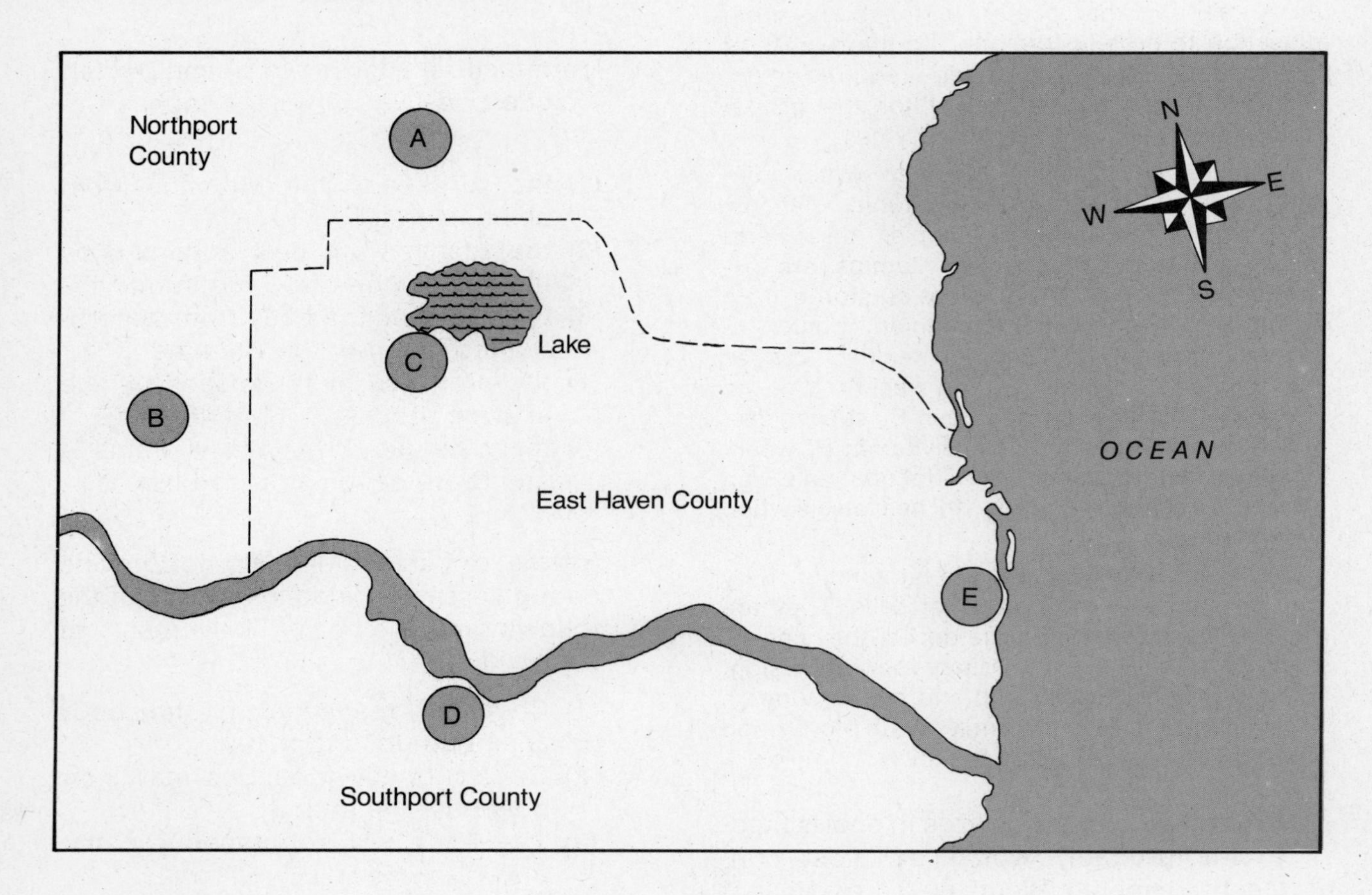

26. A person who travels from location A to location E in August experiences a 10° drop in temperature. A person who travels from location B to location E in January experiences a 10° rise in temperature. Which of the following statements offers the best explanation for these phenomena?

(1) Location A is hot in the summer; location B is cold in the winter.
(2) Location E is in a different county from locations A and B.
(3) Areas that are near the ocean are always warmer than inland areas.
(4) Cool breezes always blow off the ocean to make location E comfortable.
(5) The ocean moderates the temperature of the surrounding area.

27. A variety of forces act upon an object at any given time. An applied force pushes or pulls an object; the force of gravity draws it to the earth. Inertia tends to resist any change in the motion of an object. Friction is demonstrated when two objects rub against each other. If you are standing in a moving bus that stops suddenly, you lunge forward in the direction in which the bus had been moving. Which of the following statements best explains your motion?

(1) An applied force between your feet and the floor of the bus threw you forward.
(2) Friction between your feet and the floor of the bus stopped your motion.
(3) The inertia of your body resisted the change in motion when the bus stopped.
(4) Gravity pulled you down as the bus stopped.
(5) Motion occurs when gravity and friction work together.

Items 28 and 29 refer to the following table.

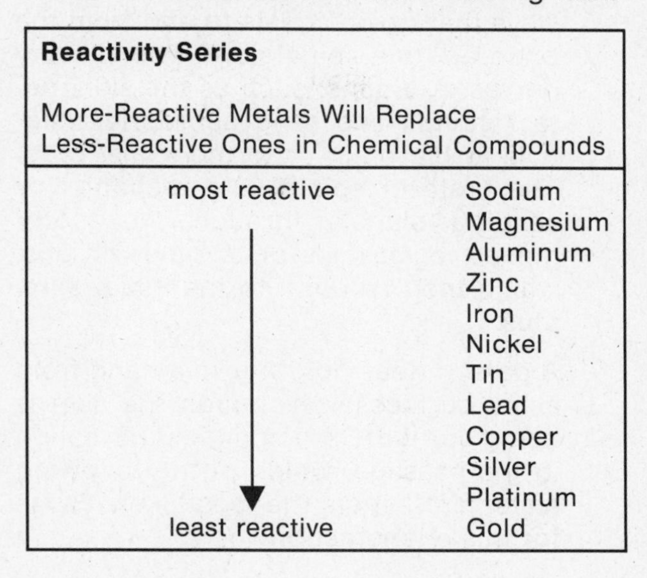

Reactivity Series

More-Reactive Metals Will Replace Less-Reactive Ones in Chemical Compounds

most reactive	Sodium
	Magnesium
	Aluminum
	Zinc
	Iron
	Nickel
	Tin
	Lead
	Copper
	Silver
	Platinum
least reactive	Gold

28. A metal bucket is lined with a compound that contains zinc. When a chemical solution is poured into the bucket, the lining begins to break down. Which of the following statements offers the best explanation for this phenomenon?

(1) The solution contained copper, which formed a new compound with the zinc.
(2) The solution contained magnesium, which replaced the zinc in the lining.
(3) The solution contained lead, which replaced the zinc in the lining.
(4) The solution contained iron, which caused the bucket to rust.
(5) The solution contained platinum, which discolored the lining.

29. Which metal is most likely to be found free and uncombined in nature?

(1) aluminum
(2) iron
(3) nickel
(4) platinum
(5) sodium

30. Throughout the body, nerves form pathways that carry signals to and from the brain and the spinal cord. The sensory nerves in organs, such as the skin, detect stimuli and send appropriate signals to the brain or to the spinal cord. The brain or spinal cord responds by sending signals through the motor nerves to muscles or to other organs, which then respond to the initial stimulus.

A person does not remove a hand from a hot surface even though the skin is being burned. Based on the information in the passage, which of the following statements offers the most likely cause for this phenomenon?

(1) The person has stronger nerves than most people.
(2) The surface is not hot enough to stimulate the nerves in the person's hand.
(3) The person's brain is not relaying the appropriate messages to the muscles in the person's hand.
(4) The burn has caused a muscular injury that makes it impossible for the person to move the arm.
(5) The person's motor nerves have not sent appropriate messages to the brain.

31. Matter on Earth is generally found in one of three states: solid, liquid, or gas. A solid has a definite shape and volume. The molecules of a solid are bound closely in place and they move very little. At the other extreme, the highly energetic molecules of a gas move independently of each other. Thus, a gas will expand to fill any container.

It can be inferred from the information given that the molecules of a liquid

(1) move faster than the molecules of a gas
(2) are less dense than the molecules of a gas
(3) differ in shape from the molecules of a gas
(4) move less freely than the molecules of a gas
(5) are like the molecules of a gas in every way

Items 32 to 36 refer to the following passage.

Soil is a mixture of minerals, decayed organic matter, water, and air. Soil can be formed in two ways. Transported soil is deposited by moving water, such as a river or runoff. For example, the soil of a flood plain is transported soil. Residual soil forms as rock weathers. The uppermost layer of a residual soil is the topsoil. Topsoil contains humus, the decayed organic matter that is formed by microorganisms.

Most soil-dwelling organisms are found in the topsoil. Under the topsoil is the subsoil, which generally is a much thicker layer. Water that percolates through the topsoil carries fine clay particles and dissolved minerals down into the subsoil. The topsoil is loose, airy, and rich in organic nutrients, but the subsoil is densely packed and contains high concentrations of minerals. Under the subsoil is partially weathered bedrock; beneath that is solid bedrock.

The physical characteristics of a soil, such as drainage and density, depend upon the size of the particles. The smallest particles are fine clay (less than .00008 inch in diameter). The diameter of silt particles ranges from .0008 inch to .002 inch, and coarse grains of sand can be from .002 inch to .08 inch across. A soil that is largely sand will drain rapidly. A soil that is mostly clay will drain very slowly because of the dense packing of its fine grains.

Most often a soil contains a mixture of particle sizes. Loam, for example, contains about equal amounts of clay, silt, and sand; it combines the advantages of finely packed clay with the open spaces of coarse grains. As a result, it has good but not excessive drainage, is fairly loose, and will have air spaces between the grains that are needed by living organisms. Plants prosper in this type of soil.

Loam can be modified by the addition of humus. Humus makes soil porous and spongy, and it adds organic nutrients. Therefore, it improves a soil's ability to sustain plant life. The decomposers that form humus grow best in moist, warm climates. This is one reason that moist, warm climates support a rich diversity of plants.

32. As a river flows down the side of a mountain, it carries particles of rock that eroded from the slopes. Eventually, these particles will form

(1) bedrock
(2) transported soil
(3) residual soil
(4) humus
(5) minerals

33. The roots of cacti, which grow in the desert, will rot if they cannot get enough air. What should be added to soil to grow healthy cacti?

(1) clay
(2) topsoil
(3) shards of bedrock
(4) sand
(5) silt

34. A farmer wants to dig a pond on his property. What soil base should he use to protect against seepage of the water into the surrounding land?

(1) clay
(2) humus
(3) loam
(4) sand
(5) silt

35. After a building was demolished, the owner of the property decided to plant a garden on the site. However, the plants did not flourish as expected—probably because the garden soil was deficient in

(1) microorganisms
(2) air
(3) water
(4) mineral particles
(5) organic matter

36. Where in the United States would you expect to find the soil that is richest in organic nutrients?

(1) the Gulf Coast
(2) the Midwest
(3) the Southwest
(4) the Great Lakes region
(5) New England

Items 37 to 41 refer to the following passage.

For matter to change from one state to another, energy must be added or removed. Adding energy in the form of heat causes the particles of matter to move faster and to spread farther apart. Removing energy, or cooling, causes the particles to slow down and to move closer together. Listed below are the processes by which matter changes states.

(1) **melting:** the process by which a solid becomes a liquid
(2) **boiling:** the process by which a liquid rapidly becomes a gas
(3) **evaporation:** the process by which a liquid slowly becomes a gas at a temperature that is below the boiling point of the liquid
(4) **condensation:** the process by which a gas changes to a liquid
(5) **freezing:** the process by which a liquid changes to a solid

Each of the following items describes a situation that refers to one of the five categories that are defined above. For each item, choose the one category that best describes the situation. Each category may be used more than once in the set of items, but no question has more than one best answer.

37. If a glass of water sits out for several days, much of the water will be gone. In time, all of the water will be gone. The water in the glass is undergoing

(1) melting
(2) boiling
(3) evaporation
(4) condensation
(5) freezing

38. The steam that rises out of a tea kettle on the stove is water that is in the gas or vapor state. The steam is formed when the liquid water inside the kettle undergoes

(1) melting
(2) boiling
(3) evaporation
(4) condensation
(5) freezing

39. As the steam hits a cooler surface, such as a shelf above a stove, droplets of water begin to collect. This process is called

(1) melting
(2) boiling
(3) evaporation
(4) condensation
(5) freezing

40. At room temperature, the metal copper is a solid. Raising the temperature of the copper to 1980° F gives the copper particles enough energy to spread apart. As a result, the copper can be poured. The copper changes states through a process called

(1) melting
(2) boiling
(3) evaporation
(4) condensation
(5) freezing

41. At 1472° F, table salt is a liquid. As the temperature drops, the salt particles slow down and the attractive forces between them cause crystals to form. The salt changes states through a process called

(1) melting
(2) boiling
(3) evaporation
(4) condensation
(5) freezing

Item 42 refers to the following figure.

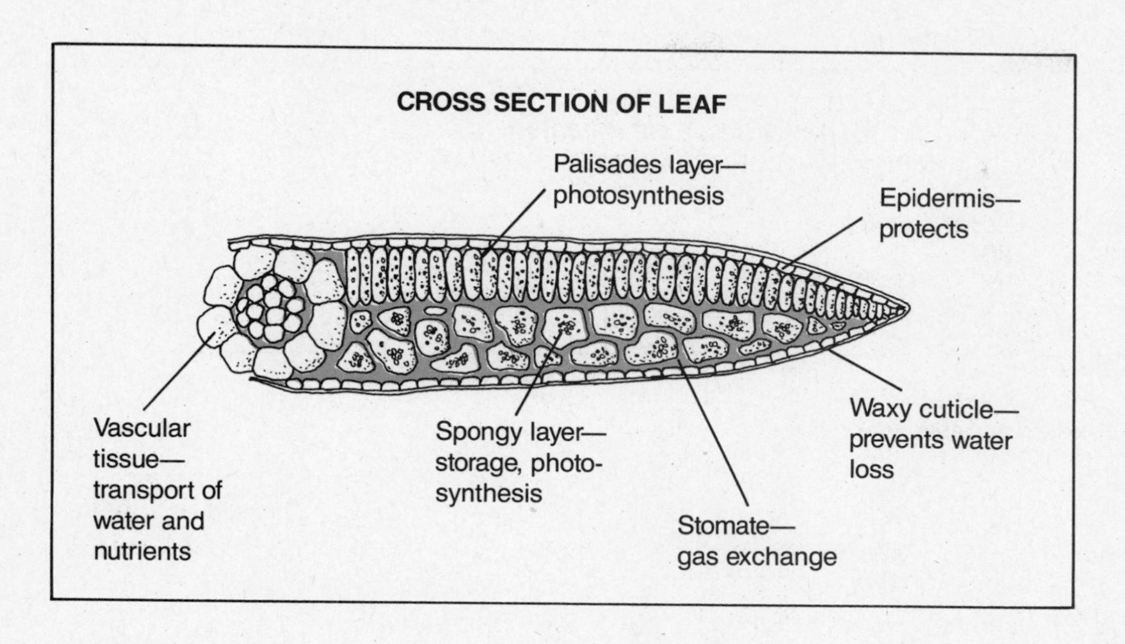

42. Which structure in the leaf of a plant serves the same function as the circulatory system in an animal?

(1) the epidermis
(2) the palisades layer
(3) the spongy layer
(4) the vascular tissue
(5) the waxy cuticle

---

43. Plants require carbon dioxide and water to form sugars by the process of photosynthesis; in that process, they produce oxygen. Other cellular processes produce carbon dioxide and water by the use of oxygen. These gases, which include water vapor, enter and leave the body of the plant through pores called **stomata.** Sunlight and warm temperatures can increase the rate of evaporation through the stomata.

Plants that grow in dry climates adapt to conserve water. According to this passage, which of the following would be an adaptation to a dry climate?

(1) not carrying out photosynthesis
(2) not having stomata
(3) reducing other cellular processes
(4) opening stomata only at night
(5) increasing the rate of evaporation

Items 44 to 48 refer to the following chart and text.

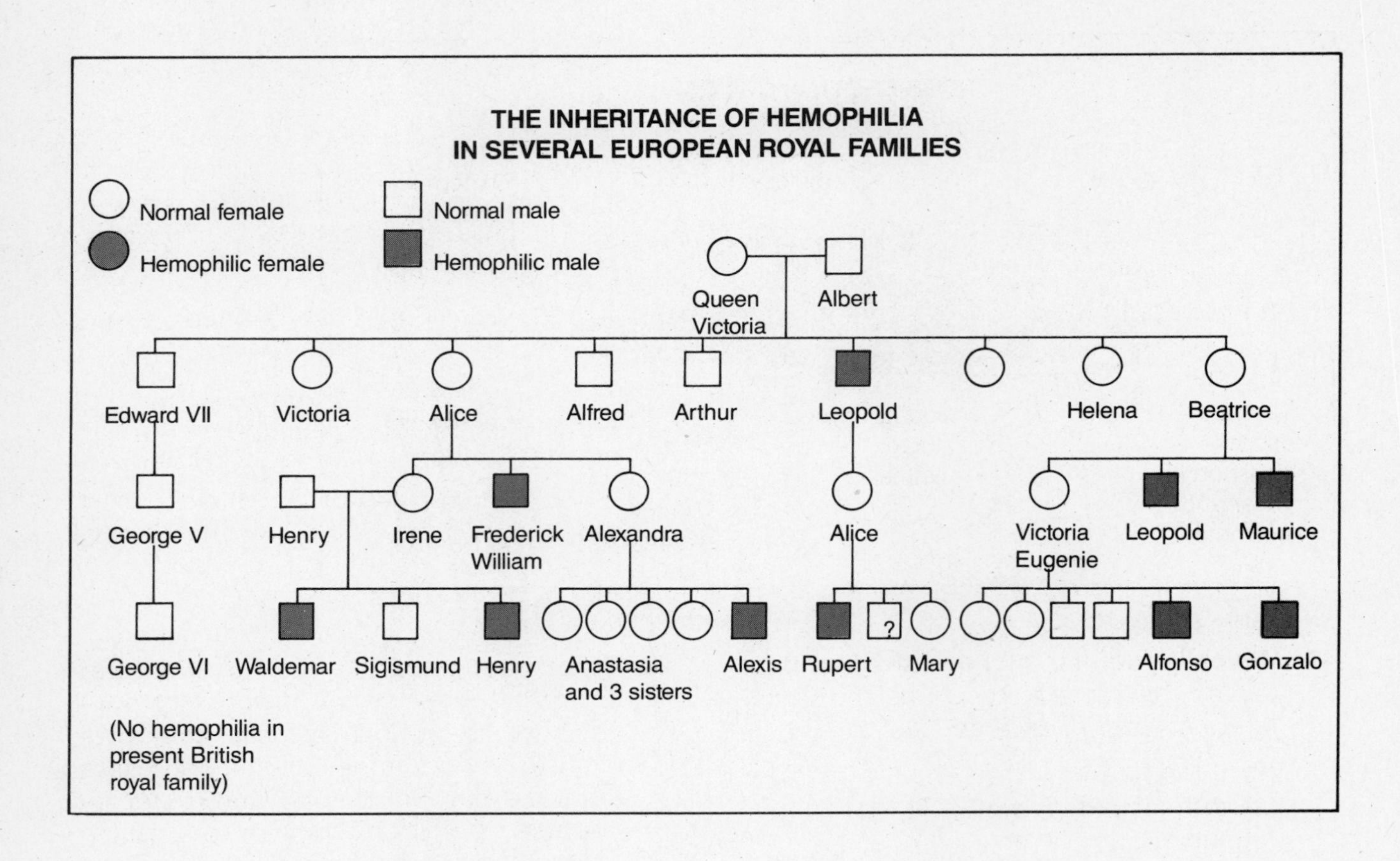

Hemophilia is a recessive, sex-linked trait that is carried by the X chromosome. Males inherit sex-linked conditions from their mothers. A female who has only one gene for hemophilia will not have the disease but will be a carrier.

44. According to the chart, which of the following statements is true?

(1) No female children of Victoria and Albert had hemophilia.
(2) No male children of Victoria and Albert had hemophilia.
(3) Anastasia and her sisters had hemophilia.
(4) There are no cases in which two brothers or two sisters had hemophilia.
(5) Hemophilia has died out in Europe.

45. A mother has hemophilia. What are the chances that her son will inherit the illness?

(1) 0%
(2) 25%
(3) 50%
(4) 66%
(5) 100%

46. What evidence is there that Queen Victoria was a carrier of hemophilia?

(1) Helena did not have hemophilia.
(2) Leopold had hemophilia.
(3) Albert did not have hemophilia.
(4) There is no hemophilia currently in the British royal family.
(5) Three generations later, hemophilia persists in Europe.

47. The royal families of Europe intermarried extensively. What evidence does this chart provide to warn against such a practice?

(1) Inbreeding causes recessive traits to appear more frequently.
(2) More children are produced through inbreeding.
(3) More male children will be born as a result of inbreeding.
(4) Hemophilia becomes more deadly through intermarriage.
(5) Hemophilia is more easily transmitted between relatives.

48. A young woman comes from a family that has a history of hemophilia, but she does not have the disease herself. Which of the following statements concerning her risk of bearing a hemophiliac child is the most accurate?

(1) There is no chance that she will bear a hemophiliac child.
(2) If the father of the child is a hemophiliac, the child may inherit the disease.
(3) She may develop hemophilia in later years; therefore, she should not bear children.
(4) There is a high probability that she will pass hemophilia on to her offspring under any circumstances.
(5) It is certain that all her sons will have hemophilia.

---

49. The *dew point* is the air temperature at which water vapor in the air will condense. As air cools overnight, it holds less moisture; therefore, some water vapor will condense as dew or frost. The dew point varies with the amount of moisture in the air; the more water vapor there is in the air, the higher the temperature at which it will begin to condense.

A gardener should be most concerned about frost damage on a night when

(1) the air is cool and damp
(2) the air is warm and moist
(3) the air is cool and dry
(4) the dew point is lower than normal
(5) there has been no rain for several days

Item 50 refers to the following figure.

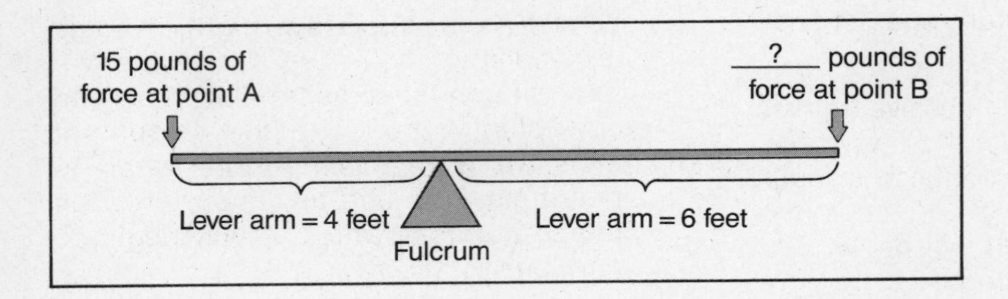

50. For a lever or a seesaw to be balanced, the product of the force and the length of the lever arm must be the same on both sides of the fulcrum. The force in pounds at point B must be

(1) 4
(2) 6
(3) 10
(4) 14
(5) 60

Items 51 to 55 refer to the following passage.

The respiratory tract is a series of passages and chambers that lead from the mouth and the nose to the lungs. Air enters the respiratory tract when the domed diaphragm muscle at the base of the rib cage flattens. This action increases the volume of the chest cavity and decreases the air pressure within it. Air is drawn in to fill the partial vacuum that is created. When the diaphragm relaxes, it pushes up, and air is expelled.

Inhaled air moves into the pharynx, or throat, a chamber that is common to both the respiratory and the digestive tracts. Swallowed food moves into the opening of the esophagus at the back of the pharynx. The food is prevented from entering the respiratory tract by a flap of tissue that blocks that passage as a person swallows. The air passes from the pharynx through the larynx, or voice box, which contains the vocal cords. These pairs of muscles, stretched across the larynx, vibrate as exhaled air passes over them, producing sound. Air next enters the trachea, or windpipe, a 5-inch tube that is held open by rings of cartilage.

At the end of the trachea, the respiratory tract divides into a right branch and a left branch, each called a **main bronchus.** These bronchi enter the spongy tissue of the lungs and divide into smaller and smaller bronchi, which terminate in about 300 million alveoli, or small air sacs. The alveoli are richly supported with blood vessels. Oxygen from the inhaled air enters the blood and is carried to the organs of the body; carbon dioxide, a waste product of metabolism, is released from the blood into the alveoli to be exhaled. For this exchange of gases to occur, the thin walls of the alveoli must remain moist.

51. A surgeon who performs a tracheotomy cuts through the windpipe to make a "shortcut" in the path of inhalation. Blockage of which of the following would be helped by a tracheotomy?

(1) the alveoli
(2) a main bronchus
(3) the diaphragm
(4) the larynx
(5) the esophagus

52. People who talk as they eat risk choking on their food. Which of the following statements offers the best explanation for such choking?

(1) Talking requires that the connection between the throat and the larynx remain unblocked.
(2) Large pieces of unchewed food may become stuck in the esophagus.
(3) The diaphragm is unable to flatten, and air cannot be drawn in.
(4) The blood supply to the alveoli may be cut off.
(5) The bronchi may collapse so that air is unable to pass through them to the alveoli.

53. In certain kinds of respiratory diseases, blood that is leaving the lungs does not have an adequate supply of oxygen. Which of the following statements offers the best explanation for this oxygen deficiency?

(1) The alveoli are dried out or damaged.
(2) There was not enough carbon dioxide in the blood.
(3) The vocal cords are vibrating.
(4) A vacuum is created.
(5) The blood supply to the alveoli is too great.

54. The suffix *-itis* means "inflammation of." Why will cough medication generally not alleviate laryngitis?

(1) The medication will not act until after it has passed the larynx.
(2) The medication will not reach the alveoli.
(3) The medication acts too slowly.
(4) The medication will enter the esophagus.
(5) Breathing will continue to aggravate the situation.

55. Certain gases, such as carbon monoxide, enter the blood more readily than oxygen, thus decreasing the amount of oxygen that can be carried by the blood. These gases are often components of air pollution, and they can build up in the atmosphere during a temperature inversion. Which of the following is appropriate advice during a temperature inversion?

(1) Sickly persons should use a humidifier or should take steam baths.
(2) All persons should lower the temperature in their homes.
(3) All persons should exercise as much as possible to increase their breathing rate.
(4) Persons who are prone to respiratory problems should stay indoors and should not exert themselves.
(5) Muscle relaxants should be taken to keep the diaphragm from becoming overstressed.

---

56. Most organisms that live in water take the oxygen that they need directly from the oxygen that is dissolved in the water. Different organisms have different levels of tolerance for low oxygen levels. When water is in contact with the air, oxygen easily dissolves in the water. Cold water can hold more oxygen than warm water can. Stagnant water often contains very little oxygen because the oxygen is used up by microorganisms and is not easily replenished.

A rapidly moving river that is located near an industrial complex is polluted periodically by hot water from a steel factory. Which of the following is the most likely effect of this practice?

(1) The amount of oxygen in the river is permanently reduced.
(2) The amount of oxygen in the river fluctuates.
(3) The water in the river becomes stagnant.
(4) The number of microorganisms in the river decreases.
(5) The amount of oxygen in the river periodically increases over the normal level.

Items 57 to 60 refer to the following information.

Radioactive decay is the spontaneous release of particles and energy from the nucleus of an atom. It is the only natural process in which matter changes from one element to another; for when an atom decays, it becomes an atom of another element.

Three types of radiation can be emitted from radioactive elements: alpha, beta, and gamma radiation. Alpha rays consist of particles that have the same mass and charge (+2) as a helium nucleus. These particles can be stopped by a piece of paper. Beta particles have the same mass and charge (−1) as an electron; they can penetrate paper but can be stopped by a thin sheet of aluminum. Gamma radiation, which is a form of electromagnetic radiation, has no charge. Gamma rays are the most penetrating form of radiation and can be stopped only by very dense materials such as lead or concrete. Alpha radiation and beta radiation are nearly always accompanied by the emission of gamma rays.

Matter changes form when it emits alpha or beta radiation. Loss of an alpha particle, for instance, converts an atom of uranium into an atom of thorium. Thorium is also radioactive, and it emits beta particles. The loss of beta particles converts the thorium atom to an atom of protactinium, which is radioactive, as well. This chain of decay continues until nonradioactive lead is formed.

Energy is released when an atom decays. Efforts to harness this natural source of energy have led scientists to develop the process of nuclear fission, in which the nucleus of an atom is split. Splitting of the nucleus is accomplished in a nuclear reactor, where neutrons are shot at the nuclei of uranium atoms. When a neutron is absorbed, the uranium nucleus splits, releasing heat energy and more neutrons. These neutrons will bombard other uranium nuclei, and the process will continue. The speed of this chain reaction is regulated in the reactor by the use of control rods, which absorb excess neutrons. The energy that is produced by nuclear fission is used to heat water to form steam. As the steam rises, it drives a turbine, which generates electricity.

57. An atom of element A changes into an atom of element B, which changes into an atom of element C, which stops changing after it finally changes into an atom of element D. Which of the following statements must be true about elements A, B, C, and D?

(1) All four elements are radioactive.
(2) Elements A, B, and C are radioactive; element D is not.
(3) Elements B, C, and D are radioactive; element A is not.
(4) One of the elements must be uranium.
(5) At least two of the elements must emit alpha particles.

58. A scientist wishes to separate a beam of radiation into alpha, beta, and gamma particles. The most effective tool for this process would be

(1) an electric field
(2) a nuclear reactor
(3) blocks of lead and concrete
(4) sheets of paper
(5) a steam generator

59. As a chain reaction begins, the control rods in a nuclear reactor fail to function. Which of the following would be the most likely result of this failure?

(1) The chain reaction would die out.
(2) Fewer neutrons would be released.
(3) The reactor would fill with water.
(4) There would be an explosion.
(5) Fewer uranium nuclei would be hit by neutrons.

60. One of the difficulties that is connected with the operation of a nuclear reactor is the method of disposal of radioactive wastes. Based on the information in the passage, which of the following statements offers the most reasonable explanation for this problem?

(1) Radioactive wastes are bulky and thus cumbersome to transport.
(2) Radioactivity from the wastes seeps through all but the densest containers.
(3) Valuable energy is lost when waste materials are removed from a reactor.
(4) The wastes must be allowed to decay before they can be disposed of properly.
(5) There is not enough land around nuclear reactors to dump radioactive wastes.

Items 61 to 63 refer to the following table.

Drugs	Medical Uses
Barbiturates (depressants)	Painkiller, sleep aid
Tranquilizers (depressants)	Sleep aid, nervousness
Penicillin (antibiotic)	Pneumonia, strep throat
Tetracycline (antibiotic)	Acne
Adrenaline (hormone)	Heart stimulant
Insulin (hormone)	Diabetes
Cortisone (hormone)	Arthritis

61. According to the table, which of the following disorders could be treated with antibiotics?

(1) nervousness
(2) acne
(3) diabetes
(4) arthritis
(5) headache

62. Which of the following drugs could be used to treat pneumonia?

(1) adrenaline
(2) insulin
(3) tranquilizer
(4) penicillin
(5) tetracycline

63. Alcohol slows the heartbeat and the rate of breathing, dilates blood vessels, and slows one's responses. Pharmaceutical companies advise against taking sleeping pills after drinking alcohol because

(1) alcohol increases urination
(2) alcohol would slow down reflexes, making it difficult to swallow
(3) alcohol may counteract the effect of the sleeping pills
(4) alcohol may interfere with the action of hormones, such as insulin and cortisone
(5) both substances slow the heartbeat and the rate of breathing, possibly to dangerously low levels

64. Communicable diseases are diseases that can be transmitted from one person to another. Influenza is an example of a communicable disease caused by a virus that affects the respiratory system. Viruses take over the body's cells, using the host cell to make more viruses. Antibiotics are not effective against viruses. However, as a patient recovers from the flu, the body produces antibodies that give it temporary immunity to the type of virus that caused the infection.

Which of the following statements is supported by information in the passage?

(1) Communicable disease-causing organisms use the body's cells to replicate themselves.
(2) Communicable diseases are caused by the influenza virus.
(3) It is rare for a virus to infect members of the same family.
(4) Antibiotics are not effective against communicable diseases.
(5) It is unlikely that a patient will have a relapse shortly after recovering from a viral infection.

Item 65 refers to the following figure.

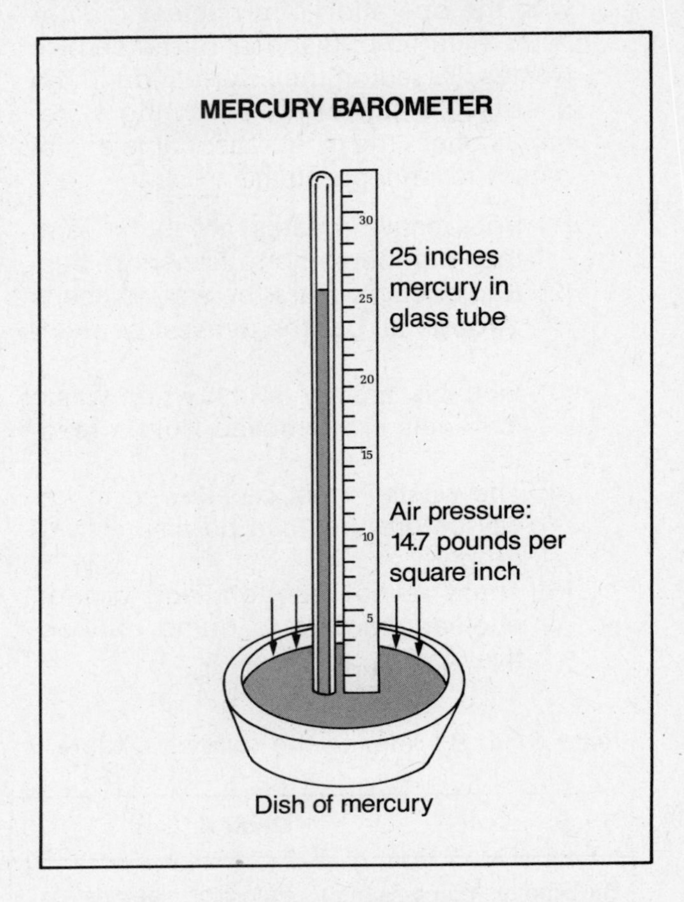

65. Why does mercury rise in the tube above the level in the dish?

(1) Air draws the mercury up the tube.
(2) Mercury can flow against gravity.
(3) The magnetic attraction of the glass pulls the mercury up the tube.
(4) Air that presses down on the dish forces the mercury up the tube.
(5) The glass contains a vacuum, which pulls up the mercury.

Item 66 refers to the following figure.

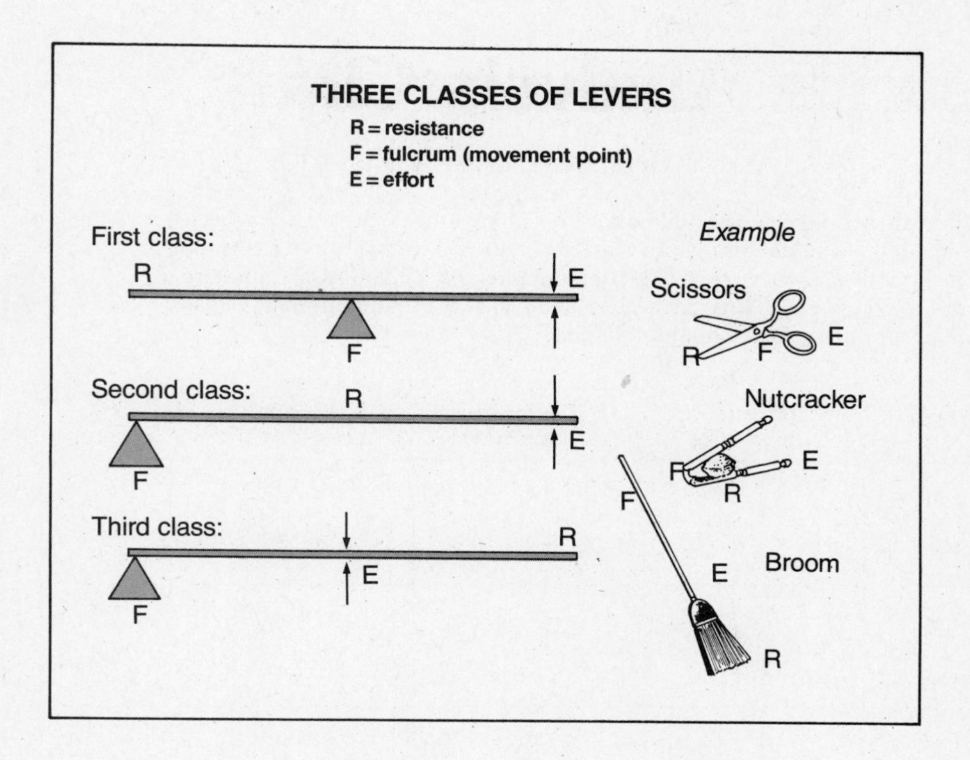

66. Based on the information in the diagram, which of the following statements is NOT true?

(1) A wheelbarrow is a second-class lever.
(2) A seesaw is a first-class lever.
(3) A second-class lever requires lifting.
(4) The fulcrum and resistance positions are reversed in first-class and second-class levers.
(5) A third-class lever can move a resistance farther than the effort moves.

Answers are on pages 276–278.

SIMULATED TEST

Performance Analysis Chart

Directions: Circle the number of each item that you got correct on the Simulated Test. Count how many items you got correct in each row; count how many items you got correct in each column. Write the number correct per row and column as the numerator in the fraction in the appropriate "Total Correct" box. (The denominators represent the total number of items in the row or column.) Write the grand total correct over the denominator, **66** at the lower right corner of the chart. (For example, if you got 56 items correct, write 56 so that the fraction reads 56/**66.**) Item numbers in color represent items based on graphic material.

Item Type	*Biology (page 83)*	*Earth Science (page 106)*	*Chemistry (page 123)*	*Physics (page 142)*	TOTAL CORRECT
Comprehension (page 31)	9, 13, 44, 61, 62	2, 10, 32	57	1, 15, 16, 50, 66	**/14**
Application (page 39)	3, 4, 5, 6 7, 14, 21, 22, 45, 51	11, 33, 34, 49	37, 38, 39, 40, 41, 58		**/20**
Analysis (page 46)	20, 23, 24, 30, 42, 43, 46, 52, 53, 54, 63	35, 56	8, 19, 28, 29, 31, 59	27, 65	**/21**
Evaluation (page 57)	18, 25, 47, 48, 55, 64	12, 26, 36	60	17	**/11**
TOTAL CORRECT	**/32**	**/12**	**/14**	**/8**	**/66**

The page numbers in parentheses indicate where in this book you can find the beginning of specific instruction about the various fields of science and about the types of questions you encountered on the Simulated Test.

GLOSSARY

acid a substance that is sour to the taste, dissolves in water, and turns litmus paper red

aerobic respiration the breaking-down of glucose into energy using oxygen

air masses pockets of air heated to the same temperature that float in the troposphere

amplitude the maximum distance a wave vibrates from its normal resting position

anaerobic respiration the breaking-down of glucose into energy in the absence of oxygen

analyze to separate a thing or idea into its individual parts

asexual reproduction process by which new cells are formed by the division of a single parent cell

asteroid belt band of small chunks of matter between Jupiter and Mars

astronomy the study of stars and the heavens

atmospheric inversion condition in which a layer of warm air traps a layer of cooler air below it

atoms the basic units of matter

axis a real or imaginary line on which an object turns

base a substance that is bitter to the taste, dissolves in water, and turns litmus paper blue

behavior the way a person or a thing acts

big bang theory the idea that the universe originated as one clump of dense matter which exploded and expanded outward

biochemistry the study of the life processes of plants and animals

biology the study of life

bond to join

calculator a machine that solves mathematical problems

capillaries the smallest fluid-carrying tubes in a plant or an animal

carbon dating the measurement of carbon-14 to establish the approximate age of a fossil

cell membrane the enclosing tissue that separates one cell from another

cell wall in plants, a tough tissue that encloses the cell membrane

cell the smallest living thing

Celsius scale thermometer on which 0° is the freezing point of water, and 100° is the boiling point

cellular respiration process by which energy is released from food

Cenozoic current era of the earth's existence

chemical change process that produces a new substance

chemical equation scientist's shorthand for what takes place during a chemical reaction

chemical properties characteristics that define how a substance will combine with other substances

chemical reaction process in which electrons are transferred or shared between atoms

chemistry the study of what substances are made of and how they change and combine

chloroplasts organelles in plants that capture light energy and use it to make the cell's food

chlorophyll the green pigment in a plant

chromosomes microscopic rod-shaped bodies that carry genes

circulatory system group of vessels and organs that carries blood through the body

cirques bowl-shaped depressions scooped out of land by a glacier's movement

classification process of identifying and naming organisms

climate the weather conditions of an area over a long period of time

cold front a mass of cool air taking the place of warmer air

compound a substance containing two or more elements that have chemically combined

computer electronic machine that performs rapid calculations or compiles data

condensation process in which a vapor or gas becomes a liquid

conditioned behavior learned response to certain stimuli

connective tissue mass of cells that connects muscles to bones

consumers organisms that live off the food products of producers

continental glaciers sheets of ice that move slowly over the land

convection currents air movement created by the rise of warmer air and the fall of cooler air

covalent bonding the joining of two atoms that share electrons equally between them

current electricity the controlled flow of electrons through an electric current

cytoplasm watery material that surrounds the nucleus of a cell

data facts and figures collected from experiments

digestive system group of organs that breaks down food into a form usable by the body's cells

dissolve to liquefy, to cause to go into solution

DNA basic chromosome material that transmits hereditary patterns

dominant the controlling gene in a chromosome, the trait that expresses itself in an organism

earthquake fractures in the earth's crust caused by strong internal pressure

earth science the study of Earth and the other planets and the forces that affect them

ecological balance state when all organisms in an ecosystem can survive

ecology the study of interdependence of organisms within an ecosystem

ecosystem an area in which plants and animals depend on each other for survival

electric current flow created by providing electrons with electric potential energy

electric potential energy power created by moving an electron away from its atom

electrolyte substance able to conduct electricity when dissolved in water

electromagnetic spectrum scale that arranges all wave frequencies

electromagnetism the generation of electricity from a magnetic field, or vice versa

electron particle of an atom that has a negative charge

electron cloud the area of an atom where an electron is likely found

elements substances made up of only one type of atom

ellipse an oval-shaped pattern; the orbit of a planet around a star

endothermic reaction chemical reaction in which energy is absorbed

energy the capacity to do work

entropy tendency of a system to achieve the most random or disorderly arrangement possible

environment the conditions and influences that surround and affect any living thing

enzyme chemical that speeds the rate of chemical reactions in the body while itself remaining unchanged

epithelial tissue mass of cells that lines and protects parts of the body

epochs units of time dividing periods

era largest unit that divides geological time

erosion process in which rocks and soil are worn down and carried away by wind and water

evaporation process in which water is turned to vapor

evolution process of genetic change that a species goes through over a long period of time

excretory system group of organs that eliminates solid and liquid wastes from the body

exothermic reaction chemcial reaction in which energy is given off

experiment test to collect data

extinct dead; gone from the earth

frequency the number of waves that pass a given point in a certain amount of time

friction the force of resistance

front formation that occurs when air masses of different temperatures meet

fault crack in the earth's crust created by internal pressure

flower plant part where gametes are manufactured

food chain the flow of energy from the sun to producers to consumers

food web a series of interconnected food chains

galaxy a star system

gametes cells with only half the normal number of chromosomes

gas state in which matter has mass but no particular shape or volume

genes various characteristics that are contained within a chromosome

genetics the study of how offspring receive traits from their parents

geological time the amount of time the earth has been in existence

glacier a large, slow-moving sheet of ice

glucose a simple form of sugar that provides cells with energy

gravity the force that any object with mass exerts on objects around it

greenhouse effect process in which the sun's energy is reflected back from water vapor and carbon dioxide in the atmosphere to keep the earth warm

habit the simplest form of learning; something that is done often

habituation behavior learned action of ignoring harmless stimuli that are repeated often

half-life amount of time it takes half the atoms of a radioactive element to decay

heredity the passing of characteristics from one generation to the next

hypothesis possible solution to a problem or a question

ice age a time period during which land masses were covered by glaciers

igneous rock substance formed when lava cools

inertia the tendency of matter to remain at its present state of motion

innate response an instinct; a behavior that is inherited rather than learned

insight learning behavior that involves putting familiar things together in new ways

insoluble substance that will not dissolve in water

ions atoms with positive or negative charges

kilogram unit of weight that contains 1000 grams

kinetic energy energy of motion

kingdom the broadest level of the classification process, contains living things that are alike in the most basic ways

laboratory a controlled environment in which experiments are carried out

learning the ability of an organism to acquire new or changed patterns of behavior based on interaction with its environment

leaves plant part that absorbs carbon dioxide and sunlight

length the distance from one point to another

light year the distance that light travels in one year

liquid state in which matter has mass and volume but no particular shape

liter the basic metric measure of volume, equal to 1.057 quarts

magnetic field force extending outward from the poles of a magnet creating an area where its magnetism can be felt

magnetism a force of push and pull, or attraction and repulsion

mass quantity of matter in an object

medium the substance through which a wave travels

meiosis division of cells in sexual reproduction that reduces the number of their chromosomes in half

Mesozoic an era of the earth's existence often called the Age of Dinosaurs

metals elements that tend to be good conductors of heat and electricity

metamorphic rock rock that has been altered by enormous heat or pressure

meter an instrument for measuring or a measurement of length equal to a bit more than three feet

microscope instrument used to view objects that are too small to be seen by the unaided eye

Milky Way the star system in which Earth is located

mitochondria organelles that produce energy for the cell

mitosis asexual division of a cell into two identical cells

mixture a material created when two or more substances are mixed together but not chemically combined

molecule the smallest particle of a substance that retains the property of that substance

moraines mounds of earth and rubble left behind by a glacier

muscle tissue mass of cells that gives the body the power of movement

muscular system group of organs and tissues that enable a body to move

mutation a sudden variation in some inheritable characteristic in a living organism

natural selection Darwin's theory of "survival of the fittest"

nebulae clouds of matter that are gradually condensed into stars and planets

nerve tissue mass of cells that makes the body aware of its environment

nervous system group of organs and nerve tissues that coordinates all functions of the body

neutralization reaction the combination of a base with an acid to form a salt and water

neutron particle in an atom that has no charge

nonmetals elements which tend to be poor conductors of heat and electricity

nuclear chain reaction process by which the split nucleus of one atom will set off the splitting of other atoms around it

nuclear energy power released by the conversion of matter during fission or fusion

nuclear fission the splitting apart of an atom

nuclear fusion the joining of two atomic nuclei to produce a new atomic configuration

nuclear reactor a device for controlling nuclear reactions

nucleus the center of a cell or atom

orbit the path that a planet follows around a star

organelles structures in a cell that control its functions

organs masses of cells organized into structures for specific functions

ozone a form of oxygen that filters ultraviolet radiation

Paleozoic an era of the earth's existence that ended 165 million years ago

periodic table chart that organizes the 109 known elements

periods units of time that divide eras

photosynthesis process in which a plant converts carbon dioxide and water into glucose and oxygen in the presence of sunlight

physical change process in which a substance's identity remains the same but the substance becomes physically different

physical properties characteristics you can identify with your senses

physics the study of different forms of energy and matter and how they behave

poles areas on a magnet from which the magnetic force extends

population the number of a given species living in a certain area

potential energy the possible energy in an object before movement

power the rate at which work is done

Pre-Cambrian Era the first four billion years of the earth's existence

precipitation process in which water returns to earth in the form of snow, rain, or hail

prevailing winds wind patterns formed by interaction between the earth's rotation and convection currents

producers organisms that create food where it did not exist before

products substances resulting from a chemical reaction

proton particle in an atom that has a positive charge

pure substance substance containing only one type of particle

radioactive decay process in which an unstable atom releases protons, neutrons, or electrons until stability is reached

radioactivity the release of energy and matter as a result of changes in the nucleus of an atom

reactants substances that begin a chemical reaction

recessive a gene that is present in a chromosome but does not show up physically in the plant or animal

recombination process of spreading new genes throughout a species by sexual reproduction

reflection property of light that allows it to bounce off objects

reflex a direct response to an external stimulus

refraction property of light that allows it to bend when passing from one medium to another

respiratory system group of organs that draws in oxygen and expels carbon dioxide

root system plant part that anchors the plant and draws food and water from the soil

rotate to spin on a center or an axis

salts chemical compounds formed by combining a base with an acid

scientific method process used to formulate a theory

scientist a person who uses facts to better understand the world

sedimentary rock substance formed when soil, sand, clay, mud, and the remains of living things are cemented together by pressure

seismic waves the shock waves that travel through the earth during an earthquake

sexual reproduction process in which new cells are formed by the union of gametes from two parents

skeletal system groups of bones and tissues that support the body and protect the internal organs

solar system a group of planets caught within a star's gravitational field

solid state in which matter has definite mass, volume, and shape

soluble substance that will dissolve in water

solute substance that has been dissolved in something else

solution mixture of a solvent and a solute

solvent substance in which something else can be dissolved

species level of classification that indicates animals that may breed with each other

states phases in which matter may exist

static electricity electricity that is the result of a buildup of charges on an object

stratosphere the upper atmosphere

subscript the number indicating how many atoms of an element are used in a chemical equation

system group of organs responsible for a vital life function

telescope instrument used to view objects in space

temperature a measure of hotness or coldness

theory a possible explanation to a problem or a question

thermodynamics study of the connection between heat energy and mechanical energy

tissue a mass of cells

trial and error a method of learning from repeated experimentation

troposphere the layer of atmosphere in which we live

unstable term that describes atoms that have trouble holding together their nuclei

vacuoles fluid-filled organelles that store food and eliminate waste from a cell

valley glaciers sheets of ice that form among mountains and move slowly downward

variable anything that can be changed

volcano an opening in the earth through which rock and lava flow

volume measure of space that an object occupies

warm front a mass of warm air taking the place of cooler air

water cycle process by which water is recycled from lakes, rivers, and oceans through evaporation, condensation, and precipitation

wavelength the distance between one crest and the next in a wave

weather the state of the air and the atmosphere

weathering the breakup of rocks caused by changing temperatures, freezing water, and plant growth

weight the force exerted on a mass by gravity

Answers and Explanations

Introduction

In the Answers and Explanations section, you will find answers to all the questions in these sections of the book:

- Lesson Exercises
- Chapter Quizzes
- Unit Tests
- The Practice Items
- The Practice Test
- The Simulated Test

You will discover that the Answers and Explanations section is a valuable study tool. It not only tells you the correct answer, but explains why each answer is correct. It also points out the reading skill that is required to answer each question successfully.

Even if you get a question right, it will help to review the explanation. The explanation will reinforce your understanding of the question and the material it was based on. Because you might have guessed a correct answer or answered correctly for the wrong reason, it can't hurt to review explanations. It might help a lot.

INSTRUCTION
UNIT I Reading Strategies

Chapter 1 A Strategy for Reading

Lesson 1

Previewing gives you an idea of how the material is organized and what it is about.

1. *Comprehension/Reading Skills.* List. There is a list of four statements about the French fries.
2. *Analysis/Reading Skills.* TRUE. It is always a good idea to look at all choices, but sometimes you can skim choices while you look for what you already know to be the answer. In Item 2, however, your preview tells you that you have to be sure you don't choose an answer that is only partially correct. Some choices, such as (1) and (3), are very similar.
3. *Comprehension/Reading Skills.* Sample wording: The material is easier to read, think about, and remember if you have an idea of what it is about and how it is organized before you start reading it.

Lesson 2

By asking questions as you read, you become more actively involved in your reading, and your comprehension improves.

1. *Application/Reading Skills.* Sample question: What are examples of some of the household materials that produce toxic fumes?
2. *Evaluation/Reading Skills.* Sample wording: "Questioning may slow your reading a bit at first, but with practice it becomes automatic. Besides, the benefit—improved understanding—is worth spending a little extra reading time."
3. *Application/Reading Skills.* Sample question: Which of these occupations would I be interested in trying?

Lesson 3

The four levels of questions are: comprehension, application, analysis, and evaluation. By identifying the level, you can adjust your thinking and reading to answer the question successfully.

1. (5) *Comprehension/Reading Skills.* Application items ask you to apply the information you are given to a new situation. Comprehension items ask you to summarize information.
2. *Application/Reading Skills.* Sample questions: Comprehension—How should a tree you buy look? Application—If you buy a small pine tree for your backyard, should you water it a lot or a little for the first few weeks after you plant it? Analysis—If a plant looks wilted, what is the probable cause? Evaluation—Is it logical to assume that all a new plant needs to grow is water?
3. *Comprehension/Reading Skills.* Sample wording: . . . Knowing the level helps you adjust your thinking and reading to the demands of the question.

Chapter 1 Quiz

1. *Comprehension/Reading Skills.* . . . You should quickly look for clues to form, such as title, labels, and layout.
2. *Analysis/Reading Skills.* Sample answer: (1) You save time because you know what you are looking for; (2) you keep yourself from getting confused by wrong answers that sound right.
3. (4) *Comprehension/Reading Skills.* This is an example of one of the shorter types of items you will find on the GED. You will never be expected to "pull information out of thin air," but sometimes you will be given only a small bit of information to work with.
4. *Evaluation/Reading Skills.* Sample wording: What can you conclude from knowing that energy is released when two things rub together? Restating information in your own words shows that you understand it.
5. *Application/Reading Skills.* Temperature in degrees F or C. Sample question: At what time of day does the person's body temperature seem usually to go down?
6. *Comprehension/Reading Skills.* The clue words *Items 6 to 9* tell you that you will need to answer four items based on the passage. The words are prominent because they are in italics, or slanted print.
7. *Comprehension/Reading Skills.* In the four items. Before reading the passage, it is a good idea to read the questions you will have to answer so that you can keep them in mind as you read.
8. *Application/Reading Skills.* Sample question: How is the decision made about which living things are plants and which are animals? The question is only partly answered in the second sentence.
9. *Comprehension/Reading Skills.* The questioning process itself is what improves comprehension. In fact, sometimes you learn more when seeking an answer that you never find, because you are thinking actively about your reading.
10. *Comprehension/Reading Skills.* 1 b; 2, d; 3, a; 4, c; 5, e. This is the suggested order. Now that you have had practice in previewing and questioning, you can sometimes skip steps or vary the order.

Chapter 2 Comprehension

Lesson 1

Putting an idea into other words helps you understand and remember it.

1. (4) *Comprehension/Reading Skills.* The length of the periods is not mentioned—Choices (1), (3), and (5). That patients feel ill during the acute period is true, but only a supporting detail.
2. *Comprehension/Reading Skills.* The incubation period is the first step in a disease when germs enter the body and multiply.
3. *Comprehension/Reading Skills.* A little more than half the people polled feel their environment will change for the worse over the next five years.

Lesson 2

To summarize ideas in written material, ask yourself: How can I express what the writer wants me to know in a sentence or two?

1. *Comprehension/Reading Skills.* FALSE. Lane begins with an "old story" that running wears out joints, then goes on to disprove it.
2. *Comprehension/Reading Skills.* Different: (1) long-distance runners versus nonrunners; (2) higher versus lower bone density. Same: (1) 41 people in each group; (2) same prevalence of osteoarthritis.
3. (2) *Comprehension/Reading Skills.* Lane shows that running has no effect on osteoarthritis, or degenerative joint disease, but strengthens bone. Although Choice (3) may be true, it is not mentioned.

Lesson 3

To identify implications, ask yourself: What can I infer about the underlying meaning of what the writer is telling me?

1. *Comprehension/Reading Skills.* TRUE. Since we know that a tick is "about the size of a pinhead," we can figure out that it must be hard to see.
2. *Comprehension/Reading Skills.* Unlikely. We are given the "clue" that the disease has caused only one death. From this, we can infer that people who get the disease are unlikely to die from it.
3. (2) *Comprehension/Reading Skills.* Because some residents want to kill the tick-carrying deer, we can infer that fewer deer might result in less Lyme disease.

Chapter 2 Quiz

1. *Application/Reading Skills.* Briefly preview the form of the diagram by glancing at captions, labels, and pictures. Ask yourself questions as you examine it more closely, such as: Where does the tick start and where does it end up over the two-year period?
2. *Comprehension/Reading Skills.* 1, c; 2, a; 3, b.
3. *Application/Reading Skills.* 1, I; 2, R; 3, S. Figuring out the level of a question you are trying to answer helps you adjust your reading and thinking. The key word *infer* tells you to "read between the lines." A reference to a specific line often tells you to "put it in other words." Choosing a title requires that you single out the main idea.
4. *Comprehension/Reading Skills.* (1) Buffalo meat has less cholesterol and fat than beef; (2) buffalo milk has more butterfat and protein than cow's milk; (3) buffalo can be more easily raised on poor land.
5. *Comprehension/Reading Skills.* Poor, wet grazing areas that produce hardly enough food for cattle to survive.
6. *Comprehension/Reading Skills.* TRUE. This is a case where it helps to add your own background knowledge to the writer's message. Generally, it is more healthful to eat food that is low in cholesterol and fat and high in protein.
7. *Comprehension/Reading Skills.* The final sentence implies that it is harder to raise cattle, which need better grazing land.
8. *Comprehension/Reading Skills.* Sample wording: First several water particles surround each protein particle. Then the small clusters clump together to form larger clusters.
9. *Comprehension/Reading Skills.* Water/protein. The diagram shows more of the small circles than the larger ones.
10. (3) *Comprehension/Reading Skills.* The circles representing the water particles are smaller than those standing for protein particles.

Chapter 3 Application

Lesson 1

Comprehension items test only your understanding of the material, but application items require you to apply that understanding to new materials.

1. *Comprehension/Reading Skills.* FALSE. You can look back in the material for a restatement, summary, or implication directly related to comprehension items. You must read for information that will help you arrive at your own answer to application items.
2. *Application/Reading Skills.* a, C; b, A; c, A; d, C. Application items can be identified because, unlike comprehension items, they refer usually to things not mentioned in the original passage or graphic.
3. (2) *Application/Reading Skills.* Farming fragile land is like winning a destructive stove because in both cases, you end up with less than you had at the start.

Lesson 2

The new type of GED item consists of a list of definitions or descriptions, followed by a set of items, each item having the same five alternatives.

1. (3) *Application/Reading Skills.* Look for a connection between the new information in the item and the old information in the passage. "Several oval units each the same . . ." sounds like "Similar cells function together," describing tissues.
2. (4) *Application/Reading Skills.* The fact that two or more kinds of tissue are combined in the heart is a clue that it is an organ.
3. (5) *Application/Reading Skills.* A quick glance back at the list of descriptions tells you that when several organs work together, you have an organ system.

Chapter 3 Quiz

1. *Evaluation/Reading Skills.* Because application items require comprehension, plus the ability to use the understanding in a new setting, application items are often harder.
2. *Comprehension/Reading Skills.* FALSE. You will find application items based on both passages and graphics.
3. *Analysis/Reading Skills.* Yes. The "new" type of items contain sets of items with the same alternatives, but each question is different. Two separate questions could have the same answer.
4. (4) *Application/Reading Skills.* The eel is living off the trout and harming it, so the relationship is parasitic. None of the other alternatives fit the definition of parasitism: The butterflies are not living off each other, the jam and dead stump are not living, and neither of the two plants is hurting the other.
5. *Application/Reading Skills.* TRUE. You need to apply what you know about the fact that tapeworms are harmful to people.
6. *Application/Reading Skills.* Sample answer: a flea on a dog.
7. (3) *Application/Reading Skills.* Only the comparison between the racing cars captures the main point of the graph: Women, although slower than men, are gaining on them.
8. *Application/Reading Skills.* Sample answer: Maybe women's marathon time would be closer by now to men's time.
9. *Application/Reading Skills.* Sample answer: Women have generally been paid less than men for doing the same work, but this inequity is decreasing.
10. *Application/Reading Skills.* TRUE. The same information can often be presented in many forms, such as in a variety of graphics or a passage.

Chapter 4 Analysis

Lesson 1

Facts are information that can be proved through measurement or observation. Opinions are interpretations of facts and cannot be proved. Hypotheses are educated guesses used to explain facts that have been observed.

1. (4) *Analysis/Reading Skills.* Studies can be done to measure whether the cheese reduces acid formation and provides calcium.
2. (5) *Analysis/Reading Skills.* The words *delicious* and *finest* signal emotional interpretations that cannot be measured and do not represent facts.
3. *Application/Reading Skills.* a, AN; b, AP; c, C; d, AN. Analysis items require comprehension, and sometimes application as well.

Lesson 2

To identify an unstated assumption, ask yourself: What must be true, although it isn't stated, for the rest of this information to be true?

1. (1) *Analysis/Reading Skills.* Don't be fooled by Choices (2) and (4), which are stated in the passage, not unstated assumptions. Because the article presents only positive aspects of the junkyards, an underlying assumption is that humans are right to intervene in correcting the problem faced by the fish.
2. *Analysis/Reading Skills.* TRUE. The writer does not choose to mention any problems, although there may well be some.
3. *Analysis/Reading Skills.* wording: Eat the fish. Perhaps the predators will now run into problems surviving because of humans' action.

Lesson 3

Supporting details are the facts or ideas that are included because they all lead to the conclusion or main point the writer is trying to make.

1. (2) *Analysis/Reading Skills.* Choices (1) and (3) are both found in the passage, but as supporting details, not conclusions. Choices (4) and (5) might be true but are not mentioned.
2. *Analysis/Reading Skills.* (1) All 100 cancer patients had the protein; (2) when the protein was found in one person thought to be healthy, it turned out he actually had cancer.
3. *Analysis/Reading Skills.* Sample wording: Only the 200 cancer victims had the marker protein in their blood.

Lesson 4

When examining several possible explanations, for each choice, ask yourself: Would this cause the change described, would it cause another change, or is it unrelated to the change?

1. (1) *Analysis/Reading Skills.* Efficient use of energy would logically explain reduced energy use. Choice (2) would contradict the findings shown. Choice (3) is true but does not provide an explanation. Choices (4) and (5) are true, but it is hard to think of a connection with lowered energy use.
2. *Analysis/Reading Skills.* Sample wording: One result may be that energy costs are being kept lower than they would be if there were more of a scarcity of energy.
3. *Application/Reading Skills.* Sample wording: . . . Is better educated about the importance of conserving energy than she used to be. Also, she has access to newer devices that are more energy-efficient.

Chapter 4 Quiz

1. *Application/Reading Skills.* 1, c; 2, b; 3, a.
2. *Comprehension/Reading Skills.* Sample wording: Your smile can give others the impression you are competent, self-confident, and charming.
3. *Analysis/Reading Skills.* All the statements are opinions, because none can be firmly proved.
4. *Analysis/Reading Skills.* The last sentence contains the conclusion. All the evidence is arranged to lead to the idea that you should go to the dentist regularly.
5. (1) *Analysis/Reading Skills.* Notice that you have to understand the phrase *death risk* before you can analyze the relationship between walking and chance of dying. Always make sure you understand a graphic or passage before you try to answer higher-level items.
6. *Analysis/Reading Skills.* Sample answer: Taking up walking is good for your health, and generally the more walking, the better.
7. (4) *Analysis/Reading Skills.*

Choices (1) and (3) are contradicted by the evidence. The author does state Choice (2). There is no evidence whether research is going on to make coal cleaner—Choice (5), although the need for such research is implied.

8. *Analysis/Reading Skills.* TRUE. The pollutants from coal can be verified by chemical analysis.
9. *Analysis/Reading Skills.* a, conclusion; b, supporting detail; c, supporting detail. The writer seems to be building a case for using coal while finding a way to prevent its harmful effects.
10. *Analysis/Reading Skills.* a, E; b, C. Because coal releases pollutants, it has a reputation for being an imperfect fuel.

Chapter 5 Evaluation

Lesson 1

To judge whether a statement is adequately supported, ask yourself: Has the writer given me enough of the right kind of information to defend what she or he is saying?

1. *Evaluation/Reading Skills.* A long period of work is *necessary* for most discoveries, but there is not enough evidence that it is *sufficient* (all that is necessary) for a discovery.
2. *Evaluation/Reading Skills.* Sample answer: It is an overgeneralization to conclude that a cup of salt will dissolve in a cup of water.
3. (5) *Evaluation/Reading Skills.* All the statements except for that in Choice (5) contain subjective descriptions (*slow, quickly, struggled*) that do not include facts.

Lesson 2

Once you are aware of how values affect beliefs, you are in a better position to judge possible bias and decide whether you want to accept what you are reading.

1. (5) *Evaluation/Reading Skills.* Oil drillers would be likely to point out all the reasons for making money from oil—Choices (1), (2), and (3)—as well as evidence that drilling won't disturb caribou—Choice (4). They would not be likely to point out that although caribou may eat near drilling sites, they may not breed there.
2. *Evaluation/Reading Skills.* a, oil driller; b, conservationist.
3. *Evaluation/Reading Skills.* Sample answer: The passage is not strongly tilted in either direction. The writer does seem to say more about what conservationists want and why; there seems to be more sympathy for conservationists.

Lesson 3

Anyone can say that something is true, but you can be more sure that what you are told are facts if the writer provides the proof, such as results of a study.

1. *Evaluation/Reading Skills.* Sample answer: A scientist might decide that the photograph was inconclusive or a fake. Don't think that just because there is documentation, something is necessarily true. The documentation itself must be convincing.
2. *Evaluation/Reading Skills.* Sample answer: Convincing documentation might be the result of a long-term study done by experts in the field who conclude from past earthquake measurements that a major earthquake is due soon.
3. (5) *Evaluation/Reading Skills.* Although the statements were presented without proof, as if they were already proven true, many people would disagree with the statements.

Lesson 4

Once you can see a misleading argument for what it is, you are less likely to be taken in by it.

1. *Evaluation/Reading Skills.* The average is found by dividing the total number of rats by the number of cages. There may be a few cages that are not crowded, but others may be.
2. (5) *Evaluation/Reading Skills.* Choices (1), (2), (3), and (4) are all generalizations based on seeing many cases of the same thing. On the other hand, you cannot generalize about all moon rock on the basis of what you observe about one type of moon rock.
3. *Evaluation/Reading Skills.* FALSE. There is not enough evidence to assume that risk is reversible.

Chapter 5 Quiz

1. *Application/Reading Skills.* b. Whenever you are asked to judge whether methods or conclusions are valid, it is an evaluation item.
2. (2) *Evaluation/Reading Skills.* Since only the Food and Drug Administration does not make money from a drug, only it has nothing to gain or lose by telling the truth about a drug.
3. *Evaluation/Reading Skills.* Sample wording: . . . Sharks contain a chemical that might help human cancer victims. (Documentation would consist of a description of research indicating that the chemical has a positive effect on treatment of cancer in humans.)
4. *Evaluation/Reading Skills.* FALSE. We are told that Dr. Luer believes that a substance in the cartilage is responsible for sharks' resistance to tumors, but we are not told why he thinks it is there and not in another part of the body.
5. *Evaluation/Reading Skills.* b. As you can see from the summary of Dr. Luer's research, summaries do not always contain enough information to allow you to evaluate the research.
6. *Analysis/Reading Skills.* Sharks/humans. Dr. Luer hopes the chemical will have the same effect on humans as on sharks.
7. *Evaluation/Reading Skills.* No, although Dr. Luer probably has evidence to support this assumption, we are not told about it. We don't have enough evidence to judge whether the reasoning is faulty.
8. *Application/Reading Skills.* a, b, and c. You should always consider the values of the writer. Even the most objective, scientific-sounding report is affected by its writer's values.
9. *Evaluation/Reading Skills.* Yes. The graph provides the evidence of where, when, and how strong each earthquake was.
10. *Evaluation/Reading Skills.* Yes. Although the quakes happened in midsummer, there is not enough evidence to say that the season *caused* the quakes.

UNIT I Test

1. *Comprehension/Reading Skills.* *Bumper crop* seems to mean a crop that is much larger than usual.
2. (2) *Comprehension/Reading Skills.* Choices (3), (4), and (5) may all be true, but the main point to which all the details lead is that trees communicate somehow. Choice (1) interprets "communicate" too loosely.
3. *Application/Reading Skills.* b. Mast fruiting refers to extra acorns and nuts (not leaves) from fruit-bearing trees.
4. *Analysis/Reading Skills.* This is a fact because it has been observed and measured.
5. *Evaluation/Reading Skills.* b. Although the farmer is probably well aware of the phenomenon, he has not kept the records of it (documentation) that the research workers have.
6. *Comprehension/Reading Skills.* Sample wording: . . . Would drop at the same speed as a copper coin in a vacuum.
7. (2) *Comprehension/Reading Skills.* Galileo and modern theorists both assume that the objects are being dropped in a vacuum.
8. *Application/Reading Skills.* The person weighing 110 pounds. Like the feather in the picture, the lighter person would fall faster because of less hypercharge working against gravity.
9. *Analysis/Reading Skills.* Hypothesis. The statement is not a proven fact, but an idea used to explain what might happen.
10. *Evaluation/Reading Skills.* To prove Galileo's theory incorrect, someone would have to create a total vacuum, drop a feather and a coin into it, and show that the times for descent were different.
11. *Comprehension/Reading Skills.* Based on context clues, melanin seems to be a substance in the skin that protects it from sunlight.
12. *Analysis/Reading Skills.* FALSE. The passage says that trying to get a tan "too fast" can cause sunburn. Thus, either short or long periods of time in the sun might or might not cause sunburn, depending on the individual involved.
13. *Application/Reading Skills.* Sample wording: Get burned when suddenly put on a hot barbecue.
14. *Evaluation/Reading Skills.* You should be careful to get a tan a little at a time so that you don't get a sunburn.
15. *Evaluation/Reading Skills.* Less. Since the article says that melanin protects the skin from sunlight, it is possible to conclude that people with more melanin are less likely to get a sunburn.

UNIT II Foundations in Science

Chapter 1 Introduction to Science

Lesson 1

The job of a scientist is to ask questions, discover new facts about nature, and to use these facts to reach a larger understanding of the world around us. A method of inquiry often used by scientists is known as the scientific method.

1. *Comprehension/Science.* Alice wants to discover whether a dry cloth or a wet cloth absorbs spills better.
2. *Comprehension/Science.* She thinks that a dry towel is better, based on her observation that it has more dry fibers for absorption.
3. *Comprehension/Science.* To carry out her experiment, she tests each cloth's absorption ability with the same amount of water.
4. *Analysis/Science.* The conclusion is a dry cloth is more absorbent than a wet cloth.
5. *Application/Science.* A partially dry cloth would have an absorbency in between that of the wet cloth and the dry cloth.

Lesson 2

Science came about as people's desire to know and to explain the environment developed into various fields of scientific study.

1. *Analysis/Science.* Science evolved gradually as people observed, discovered, and drew conclusions about what they found out. There was never a time when someone said, "Let's begin the study of biology, physics, etc."
2. *Comprehension/Science.* Ptolemy practiced astronomy, the study of the planets and the stars.
3. *Evaluation/Science.* Equipment has a great effect on scientific knowledge. Before the microscope, most cells could not be seen with the naked eye so they were difficult even to discover, much less study. Today's sophisticated instruments allow scientists to observe and study particles, etc., that would have been impossible to study even a hundred years ago.

Lesson 3

The major fields of science are biology, earth science, chemistry, and physics. These categories are not rigid because they overlap at times—a biochemist, for example, must know some biology and chemistry.

1. (4) *Application/Science.* Electromagnetic waves are a form of energy.
2. (1) *Application/Science.* Animals are living things.
3. (2) *Application/Science.* Volcanoes are part of changes in and on Earth.
4. (3) *Application/Science.* Knowing water's composition is part of the study of what substances are made of.

Lesson 4

An experiment is one of the chief ways of learning scientific principles and generalizations. Experiments are designed so that all properties remain constant except for the variable(s) under study.

1. *Comprehension/Science.* The plant pot, the plastic bag, and the soil remained the same both nights.
2. *Comprehension/Science.* The presence or absence of the plant was the variable on the different nights.
3. *Analysis/Science.* No, because he hadn't tested to see what would happen if he set up a similar experiment without the plant in the bag.
4. *Analysis/Science.* Yes, because he could compare the setup without the plant with the one that had the plant.
5. (2) *Evaluation/Science.* If the bag is not sealed tightly around the plant, water from the air could condense on the inside of the bag, making it difficult to prove where the water came from.

Lesson 5

Scientists use accurate and reliable tools such as telescopes, microscopes, clocks, meters, rulers, calculators, computers, and so forth. To measure, scientists all over the world use the metric system.

1. (4) *Application/Science.* Although scientists may use almost anything in the course of their work, a typical measuring cup designed for kitchen use would be too inexact for most scientific purposes.
2. (5) *Comprehension/Science.* A kilometer has the greatest length because it equals 100 meters.
3. *Comprehension/Science.* A liter is slightly larger than one quart, so two quarts would be greater than a liter.
4. *Comprehension/Science.* The rock has the least mass, because one-half gram is less than one gram.
5. *Application/Science.* Yes, because 100°C is the boiling point of water.

Chapter 1 Quiz

1. *Comprehension/Science.* White, because the graph shows a depth of 0 centimeters.
2. *Comprehension/Science.* Black, because the graph shows a depth of 9 centimeters.
3. *Analysis/Science.* The different colored cloths are the variable, because everything else stayed the same.
4. *Analysis/Science.* The experiment is the kind where a scientist measures a variable under different conditions; in this case the color of the cloth was what differed.
5. *Analysis/Science.* The black cloth

sank because it absorbed the most heat.

6. *Evaluation/Science.* Light or white clothing would be best because it absorbs the least heat and thus would keep you coolest.
7. (1) *Application/Science.* Milk takes up space.
8. (3) *Application/Science.* The amount of matter in you stays the same anywhere.
9. (4) *Application/Science.* Degrees Celsius measure the hot and cold.
10. (2) *Application/Science.* Length is a measure of distance from one point to another.

Chapter 2 Biology

Lesson 1

The cell is the smallest unit that performs all the fundamental functions of life—growth, reproduction, use of energy, and response to stimuli. The three basic parts of the cell are the cell membrane, the nucleus, and the cytoplasm.

1. *Comprehension/Biology.* A snowball grows as it rolls by the addition of material (snow) from outside. A tree grows by producing new cells.
2. *Comprehension/Biology.* FALSE. The four basic life processes are common to all living organisms.
3. *Application/Biology.* A secretary's function is similar to that of a cell membrane, which controls the flow of material into and out of the cell.
4. *Evaluation/Biology.* A single-celled organism is self-contained and self-sufficient. It must obtain its own food and oxygen. A cell in a multicelled organism is specialized; many cells rely on other specialized cells to provide them with food and to carry away wastes.
5. *Analysis/Biology.* As the drawing shows, nerve cells are elongated and have branches. That shape helps them relay messages between distant points.

Lesson 2

Cells obtain their energy through respiration during which glucose is broken down through a series of chemical reactions that release the energy for the cell's use. Most organisms obtain their glucose from food. Plants make their own glucose during photosynthesis.

1. *Comprehension/Biology.* Cells must obtain food, air, and water. The cell gets glucose from food and gets oxygen from air and water. A plant cell, however, makes its own glucose; instead of food, it requires sunlight. Instead of oxygen, it requires carbon dioxide.
2. *Analysis/Biology.* The plants must be near the surface in order to receive enough sunlight for photosynthesis. Below several feet of water there is not enough sunlight for the process to occur.
3. *Application/Biology.* Green plants release oxygen into the atmosphere during photosynthesis, and remove carbon dioxide.
4. *Analysis/Biology.* The two processes are alike in that both involve the conversion of carbon dioxide into fixed carbon. They are different in the source of energy used, and the organisms involved.
5. (5) *Evaluation/Biology.* Both processes support a variety of life, as shown by the arrows. There is no evidence for choices (1) and (2). Choice (3) is incorrect because the bacteria are doing the carbon-fixing job of the green plants. Choice (4) is incorrect because the arrows show that the fish feed on the snails, and the snails feed on the worms.

Lesson 3

Living things reproduce either through sexual reproduction or asexual reproduction. Single-celled organisms, and cells in a multicelled organism, reproduce through mitosis, in which the cell divides in half. Multicelled organisms reproduce sexually through meiosis, in which two gametes combine to form a new organism.

1. (2) *Application/Biology.* Asexual reproduction is achieved by mitosis, which is performed by single-celled organisms. All the other alternatives are multicelled organisms, which reproduce sexually.
2. *Analysis/Biology.* Because it is part of sexual reproduction, the egg requires a male gamete to fertilize it. If the egg is not fertilized, it will never develop into a chick, and consequently will not hatch.
3. *Comprehension/Biology.* Budding is asexual reproduction because the parent organism divides into two parts.
4. *Analysis/Biology.* A bud has the same number of chromosomes and thus is exactly like the parent organism. A gamete has half the chromosomes of the parent and thus is only partly like the parent.

Lesson 4

Characteristics are passed on through the genes, which are found in the chromosomes. If a gene is dominant, the characteristic will appear in the offspring. If the gene is recessive, the characteristic will not appear in the offspring, unless two recessive genes are present. Recessive genes will remain in the offspring's chromosomes.

1. *Comprehension/Biology.* When a gamete is formed during meiosis, the parent cell, which contains two genes for every characteristic, splits apart. Each of the new gametes formed receives half of the chromosomes, and therefore only one gene for each characteristic. When the male gamete and the female gamete combine during reproduction, they combine their genes into one set—two genes for each characteristic.
2. *Application/Biology.*

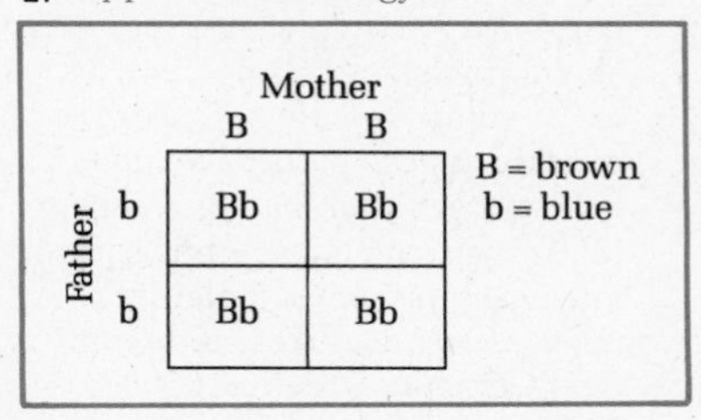

3. *Analysis/Biology.* There are two chances in four that their children will have blue eyes. (See figure.)

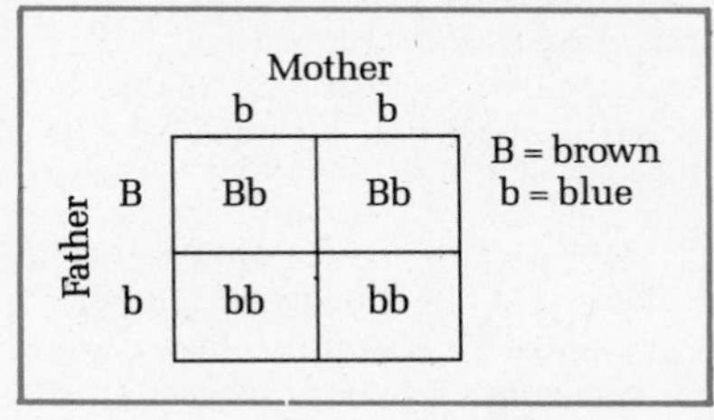

4. (2) *Analysis/Biology.* According to the figure, types A and B express themselves in every combination. Type O, however, is expressed only when paired with another type O gene, and thus is recessive.

5. *Evaluation/Biology.* Knowledge of genetics allows us to select desired traits in plants and animals, increasing the efficiency of modern agriculture. For example, plants may be selected for their disease resistance, with the result that more live to be harvested.

Lesson 5

The cells of complex organisms are specialized to perform specific tasks. Specialized cells are themselves organized into structures called tissues, which contain only the same type of cells. Various kinds of tissues are organized into organs, and the organs are organized into systems. All the systems together make up the body of the organism.

1. *Comprehension/Biology.* The highest to the lowest form of organization is system, organ, tissue, and cell.
2. *Comprehension/Biology.* A system provides the whole body with a vital life function, such as digestion, respiration, excretion, etc.
3. *Application/Biology.* The circulatory system transports food, oxygen, and chemicals to the body's cells, as the capillaries do in a plant.
4. *Analysis/Biology.* A plant makes its own food, so it doesn't need to digest large food molecules into a usable form as animals do.
5. (5) *Evaluation/Biology.* The virus directly attacks parts of the nervous system, eliminating choice (3). Because motor neurons control movement, systems containing muscles are also likely to be indirectly affected, eliminating choices (1), (2), and (4).

Lesson 6

Scientists distinguish between life forms by classification. The classification system provides many levels on which an organism can be classified.

1. *Application/Biology.* Human beings belong in the kingdom Animalia because we are multicelled, cannot make our own food, and eat other organisms for food.
2. *Analysis/Biology.* Some single-celled organisms have no nuclei, which plants and animal cells always have. Even though some single-celled organisms use photosynthesis (like plants) or seek their own food (like animals), an organism must be multicelled in order to be classified as a plant or an animal.
3. *Application/Biology.* A dog and a cat are not considered members of the same species because they cannot breed with each other.

Lesson 7

Organisms depend on each other and interact with the living and nonliving things around them. In an ecosystem, each organism provides something that the others need.

1. *Comprehension/Biology.* Larger fish depend on smaller fish and plants for food. Plants, in turn, depend on the nutrients supplied by decaying fish.
2. *Evaluation/Biology.* The dam would cause more damage because it would affect every creature in the environment. Such large changes are difficult for the natural environment to overcome. However, when an imbalance occurs in the size of a population, it is usually possible for the environment to reach a new balance. The overpopulation of a species of fish would eventually cause some of the fish to die of starvation until a new balance was achieved.
3. (2) *Evaluation/Biology.* Plowing up the grass would cause the hippos to seek food elsewhere. The other plants and animals depend on the hippos for food, either directly or indirectly. Choice (1) is incorrect because there are other predators besides people. Choices (3), (4), and (5) are also incorrect because they would create only minor imbalances in the ecosystem.

Lesson 8

Plants produce food energy by photosynthesis. Animals get energy by eating plants, or by eating organisms that eat plants.

1. *Comprehension/Biology.* This is a food web because it describes interconnected food chains.
2. *Analysis/Biology.* At each level, the energy available is about 10 percent of the energy at the level below it; 90 percent of the energy is lost from one level to the next higher level.
3. (2) *Application/Biology.* The food chain shows that grasses get energy from the sun and become food for the steer, which provides steak for the human being.
4. *Application/Biology.* More mice would be found in an ecosystem, because mice are at a lower level on the food chain. Less energy is available at each level of a food chain; thus the population of each species decreases as one moves up the chain.

Lesson 9

The way living things respond to their environment is called behavior. In general, the more complex an organism is, the more complex type of behavior it is able to exhibit.

1. *Comprehension/Biology.* Answers may vary. Examples of habit behavior are driving a car, dressing, talking, typing, dancing, and so on.
2. *Application/Biology.* Answers (a) and (d) are examples of reflex behavior. Answers (b) and (f) are examples of innate response, or instinct. Answers (c) and (e) are examples of learning behavior.
3. *Analysis/Biology.* Language puts two very different things together in a new way. It uses symbols to indicate objects, actions or ideas that may have nothing in common with the symbol. The word "horse," for instance, does not look like a horse or sound like a horse, yet anyone who speaks English knows the exact thing that the word indicates. Such an ability to connect an abstract symbol with a concrete thing is an example of insight learning.

Lesson 10

Life forms change by adapting to new conditions and by passing adaptations on to their offspring. Such changes are part of the process of evolution. The two main mechanisms of evolution are genetic variation and natural selection.

1. (1) *Analysis/Biology.* Harmful changes caused by the activities of humans often occur so quickly that the organisms do not have time to adapt, and the natural evolutionary process cannot occur. Evolution is a process of gradual change.
2. *Analysis/Biology.* Some insects affected by an insecticide are able to withstand the effects of the pesticide. Since these insects are the ones that survive, they pass the immunities on to their offspring. With

each succeeding generation, more and more insects are immune. After a time, all the insects are immune to the insecticide.

3. *Evaluation/Biology.* Vaccinations allow many people to live who would not survive by natural selection. Thus the weaker traits in the species continue to be passed on and do not disappear.

Chapter 2 Quiz

1. *Comprehension/Biology.* An increase in phosphorus leads to an increase in algae, an increase in bacteria, a decrease in oxygen level, and finally, a decrease in fish.
2. (2) *Evaluation/Biology.* Photosynthesizing organisms require light to make food. The fact that algae die from lack of light indicates that they require light for some process, most likely photosynthesis.
3. *Comprehension/Biology.* No, because a low fever is not usually dangerous and may even help to fight the disease.
4. (1) *Application/Biology.* The salt is passing from an area of strong concentration into an area of low concentration—the less salty cucumber.
5. (5) *Application/Biology.* The liver cells are taking digested food (amino acids) and putting them together to construct a new substance (protein).
6. (3) *Analysis/Biology.* Assuming that these two leaves are typical, the main difference is that the simple leaf consists of one leaf, while the compound leaf consists of two or more leaflets. Choices (1), (4), and (5) discuss minor differences that could vary among trees. There is no basis to choose choice (2).
7. *Application/Biology.* The central office of a large corporation is similar to the nervous system of the human body, because the nervous system controls and coordinates all the body's activities.
8. (4) *Analysis/Biology.* As long as one parent fly has brown spots, which are dominant, there is no way for all the offspring to have red spots (see figure). At most, there is a 50% chance that the offspring will have red spots, assuming that the brown-spotted fly has one red gene. If the brown-spotted fly has two brown genes, all the offspring will have brown spots. Choice (1) could be a true statement, although it does not have to be true; choice (2) must be true, or the fly would not have red spots.

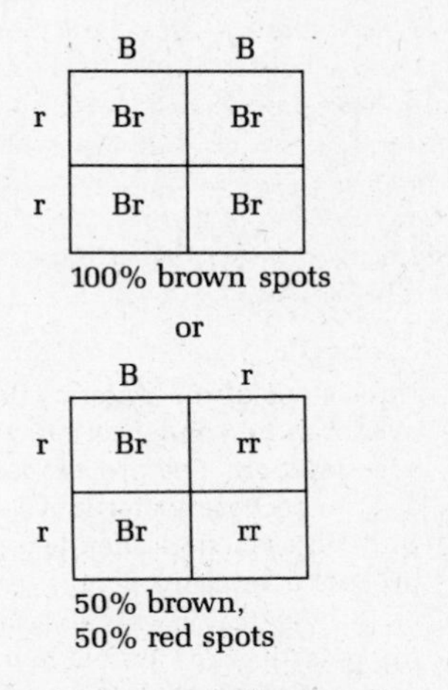

9. *Analysis/Biology.* If the snake population were drastically reduced, the frog population would probably increase, since there would be one less frog predator. If the frog population increased, the cricket population would probably decrease, since more cricket predators would exist. The hawk population might decrease as well, if hawks depended on snakes as a major food source.
10. *Evaluation/Biology.* The statement is incorrect, because the grasses in the figure depend on sunlight for energy. According to the figure, sunlight is the ultimate source of energy for the living things depicted.

Chapter 3 Earth Science

Lesson 1

Earth orbits the sun along with eight other planets in the solar system. The solar system is located on the edge of the Milky Way galaxy, which is only one of millions of galaxies in the universe.

1. *Application/Earth Science.* The Southern Hemisphere experiences the opposite season from the Northern Hemisphere, so it would be fall.
2. *Comprehension/Earth Science.* FALSE; all the stars that can be seen with the unaided eye—without a telescope—are in the Milky Way galaxy.
3. *Analysis/Earth Science.* The moon orbits Earth because it is held by Earth's gravity. (Scientists consider the moon a satellite of Earth.)
4. *Application/Earth Science.* Because of the way the Earth turns, Denver is two hours behind Washington, D.C. Thus, you should set your watch back to noon in order to have the correct local time when you land in Denver.

Lesson 2

Earth was formed by clouds of matter and gas in space that were condensed by gravity into a ball of molten metals and gases. The planet slowly cooled to become the planet we know today.

1. *Evaluation/Earth Science.* Earth is believed to have formed about 4.6 billion years ago, but the oldest rocks that can be dated are several hundred million years younger than that. Thus the rocks must have existed for a time in a state that cannot be dated—the molten form.
2. (4) *Application/Earth Science.* The lesson describes how heavy, molten metals remained in Earth's center, while lighter gases rose to the surface. Though rock, in choice (3), is heavy, it is not made of pure metal, so it is not likely to be as heavy as iron.
3. *Analysis/Earth Science.* The Precambrian is the earliest era, and thus the most time has passed, during which many fossils could have been destroyed. In addition, the earliest life forms were a type of bacteria, a single-celled organism with no hard parts. The chances of finding fossils of such organisms are small, although some do exist.

Lesson 3

Earth has a hard, outer crust that encloses three layers: the mantle, a liquid outer core, and a solid inner core. The rocks that make up the crust are of three types: igneous rock, sedimentary rock, and metamorphic rock.

1. *Application/Earth Science.* The statement is true, because high temperatures and pressure have prohibited anyone from ever exploring the interior of Earth. Most of what scientists know has come from the indirect evidence of seismic waves.
2. *Application/Earth Science.* A sedimentary rock, such as sandstone, might be forced beneath Earth's surface and subjected to high temperatures and pressure, and changed into metamorphic rock. The metamorphic rock might then melt, and eventually cool and harden into igneous rock. This is, in fact, a process that provides volcanoes with much of their lava.
3. (2) *Comprehension/Earth Science.* The other processes all take place over long periods of time, so they would not cause a sudden change.

Lesson 4

Two main types of changes occur on Earth: surface changes and subsurface changes. Examples of subsurface changes include continental drift, earthquakes and volcanoes. Examples of surface changes include weathering and erosion.

1. (1) *Evaluation/Earth Science.* The coming together of plates can cause the formation of mountain ranges, whether on land or underwater. Choice (2) is perhaps an effect of melting icecaps, but not of plate movement. Choice (3) and choice (4) are true statements, but they do not relate to the information given. Choice (5) contradicts the correct answer.
2. *Comprehension/Earth Science.* Subsurface changes cause surface changes by pushing up, folding, or breaking Earth's crust. Continental drift can cause the formation of mountain ranges. Volcanoes can create mountains or islands with lava flow. Earthquakes split the earth and cause considerable damage on Earth's surface.
3. *Application/Earth Science.* Erosion and weathering can break down the rocks that form mountains and move them. In this way, mountains that may once have been very high, such as the Appalachian Mountains in the United States, are slowly eroded into smaller ranges.
4. *Analysis/Earth Science.* Erosion from wind and water breaks down and carries away soil particles. Planting a ground cover helps keep the soil in place, because the plant's roots help to anchor the soil.

Lesson 5

Water covers 70% of Earth's surface. Most life on Earth requires water to live, so without water, life as we know it would be impossible. Water in oceans, lakes, and rivers supports many kinds of life directly. Water also affects life indirectly, by creating surface changes through erosion. In addition, water plays an important part in weather.

1. *Comprehension/Earth Science.* The three steps of the water cycle are evaporation, condensation, and precipitation. The process is called a cycle because water moves through each step, then begins the process over again.
2. (5) *Analysis/Earth Science.* Because the polar icecaps are made up of fresh water, their melting would increase the total amount of available water, not cause fresh water to become scarce. The other choices are all possible consequences of increasing the amount of fresh water entering the ocean.
3. *Comprehension/Earth Science.* The two points where the sun and moon's pulls are aligned cause the highest tides. These are positions 1 and 3.
4. *Application/Earth Science.* The gravitational pull of the moon causes high tide. If the moon rises fifty minutes later each day, tides must occur fifty minutes later each day. Fifty minutes after 7:50 A.M. is 8:40 A.M., the time you would need to set sail in order to leave at high tide.

Lesson 6

The atmosphere provides the gases that all living things need in order to carry out life processes. It protects Earth's surface from harmful elements and radiation from outer space.

1. *Comprehension/Earth Science.* The ozone layer protects Earth from the sun's harmful ultraviolet rays. A thinning of the ozone layer would be hazardous to people because they would be exposed to too much ultraviolet radiation, leading to burns, skin cancer, and even blindness.
2. *Analysis/Earth Science.* Gravity holds the gases in the atmosphere to Earth. The force of gravity is strongest close to Earth's surface. Therefore, "heavy" gases such as nitrogen and oxygen are found near Earth's surface, while "light" gases such as hydrogen and helium rise high into the atmosphere. In addition, more gases are held close to the earth. The proportion of gas to empty space drops off the farther you go from Earth.
3. *Comprehension/Earth Science.* Oxygen is taken in by animals, who give off carbon dioxide. Plants take in carbon dioxide, and give off oxygen. Animal and plant remains are broken down by decomposers, which give off carbon dioxide. Thus, there is a constant give-and-take of these two gases between plants, animals, and decomposers.
4. *Analysis/Earth Science.* It is Earth's forests which remove much of the atmosphere's carbon dioxide. If these were destroyed, there would be an oxygen shortage and a surplus of carbon dioxide. Extra carbon dioxide would increase the greenhouse effect, since carbon dioxide is one of the gases that trap the sun's energy. An increase in the greenhouse effect would lead to an increase in temperatures.

Lesson 7

The main factors that affect climate are a region's altitude, proximity to water, and location on the globe. The main causes of weather are unequal heating in the atmosphere, air pressure, and the rotation of the planet.

1. *Comprehension/Earth Science.* When warm and cold air masses meet, a front is formed. Weather is usually cloudy and rainy along a front.
2. (5) *Analysis/Earth Science.* Warm air rises as cold air moves in underneath it, indicating that it is less dense. Choice (3) is incorrect because both types of fronts are accompanied by a change in weather. Choices (1), (2), and (4) are not referred to in the lesson.
3. (4) *Application/Earth Science.* The principal movement is upward because hot air rises. (This area is called the doldrums because winds tend to be light and erratic due to the upward movement of air.) The winds may then be deflected west-

ward, but their principal movement is upward.

4. *Analysis/Earth Science.* Precipitation is heavier along a cold front because water condenses out of the air in a shorter period of time, due to the greater temperature differences.

Lesson 8

Solar and geothermal energy are the ultimate sources of the energy that shapes this planet. Solar energy powers the weather, which in turn contributes to erosion. Geothermal energy creates plate movement, earthquakes, and volcanoes. Both are used by life on Earth as energy sources as well.

1. *Application/Earth Science.* Renewable sources of energy are geothermal energy, solar energy, wind energy, and flowing water. These sources are renewable because, no matter how much of their energy is used, there will always be more to replace it. The nonrenewable sources are fossil fuels, such as oil and coal. When these are used up, they must be replaced by other sources.
2. (3) *Evaluation/Earth Science.* Electricity may be generated by harnessing wind, water, solar, and nuclear energy, as well as by burning fossil fuels. Choice (1) is incorrect because Earth has usable chemical energy available from plants that other planets lacking life do not have. The other choices are all supported by statements made in the lesson.
3. (3) *Analysis/Earth Science.* Igneous rock is formed by volcanoes (see Lesson 3) and thus would be hottest where the activity is most recent.

Chapter 3 Quiz

1. *Comprehension/Earth Science.* Due to the tilt of Earth's axis, the Northern hemisphere is tilted toward the sun for part of the year and away from the sun for part of the year. Winter occurs when the hemisphere is tilted away from the sun.
2. *Comprehension/Earth Science.* Igneous rocks are formed from hot, molten rock deep beneath Earth's surface. Metamorphic rocks are changed from one type of rock into another by heat and pressure beneath Earth's surface.
3. *Analysis/Earth Science.* Warm weather would cause more water than usual to evaporate from ocean and land sources. Warm air temperatures would cause less water vapor than usual to condense and to fall back to Earth as rain. The result would be dwindling water supplies and little rainfall—probably a drought.
4. *Application/Earth Science.* (a) climate; (b) weather; (c) weather; (d) climate
5. (4) *Evaluation/Earth Science.* A is evidence that conditions exist that favor the greenhouse effect. B is further evidence, since Venus is farther away from the sun than Mercury, and thus logic would dictate that Venus should have lower average temperatures than Mercury, not higher. C is not evidence, since relative distance to the sun has nothing to do with the greenhouse effect as defined.
6. (2) *Evaluation/Earth Science.* Both oxygen and carbon dioxide are present in the Martian atmosphere, which could be indirect evidence that animals exist on Mars.
7. *Analysis/Earth Science.* If present, ammonia must exist in a state other than gas on Uranus and Neptune. Since the temperatures of these planets is lower than that of Saturn and Jupiter, ammonia is probably in the liquid or frozen form on Uranus and Neptune.
8. *Application/Earth Science.* The satellite would travel into outer space in a straight line. It would not stop until it came under the influence of a body in space and was pulled toward it by that body's gravitational field.
9. (1) *Analysis/Earth Science.* Earth completes a rotation from west to east every 24 hours, so a plane traveling south would end up somewhere west of its destination. Choices (2) and (5) are false because Earth rotates toward the east, not the west. Convection currents could push the plane up, down, or sideways, depending on where the plane went through them, eliminating choice (3). Wind resistance would tend to push the plane backward, not westward, eliminating choice (4).
10. *Application/Earth Science.* The length of a day is determined by the amount of time it takes for Earth to make one complete rotation—currently around 24 hours. Since Earth's rotation is gradually being slowed, the amount of time to complete a rotation is increasing, causing the day to lengthen over many years.

Chapter 4 Chemistry

Lesson 1

Matter is anything that has mass and volume. All matter has physical and chemical properties that differ according to the substances involved.

1. *Application/Chemistry.* Sublimation is a physical change because it doesn't result in new kinds of matter. Whether it is in the form of frost or vapor, the substance is still water.
2. *Analysis/Chemistry.* A liquid has definite volume, so that in order to inflate a tire, much more liquid would have to be used, making for a relatively heavy tire. A gas, on the other hand, takes on the shape and the size of its container, making it a much lighter substance with which to inflate the tire, and thus increasing fuel efficiency.
3. (3) *Evaluation/Chemistry.* The fact that the leaves can no longer make food seems to indicate that the identity of a substance in them (chlorophyll) has been changed, indicating a chemical change. Choices (1) and (4) are not changes in the leaves themselves. Choice (2) is a physical change. Choice (5) could trigger a physical change or a chemical change.

Lesson 2

The basic unit of matter is the atom, which is formed of three different particles—protons, neutrons, and electrons. Atoms share or transfer electrons to form larger units of matter.

1. *Comprehension/Chemistry.* Hydrogen and oxygen are atoms, because they have different properties individually than they have when bonded.
2. *Application/Chemistry.* The formation of table salt is an example of ionic bonding, because the sodium

atom gives an electron to the chlorine atom.

3. (1) *Evaluation/Chemistry.* The chlorine atom gaining an electron is a reduction reaction, and the sodium atom losing an electron is an oxidation reaction. Choice (2) is irrelevant information. Choices (3) and (4) could be true whether or not an oxidation-reduction reaction was involved.
4. (3) *Analysis/Chemistry.* An ion is either positively or negatively charged, not neutral. The other statements are true.

Lesson 3

The periodic table lists atoms by symbol and atomic number. The atomic number tells how many protons are in each atom of an element. Elements that have similar properties are arranged in groups. In addition, all the metals are on the left side of the table and most of the nonmetals are on the right side of the table.

1. *Comprehension/Chemistry.* Hydrogen, carbon, and oxygen are all listed in the periodic table and thus are elements themselves. Vinegar is a compound, because it is a combination of three elements.
2. (1) *Analysis/Chemistry.* Water and hydrogen have at least one proton, one neutron, one atom (hydrogen), and one electron in common. They do not have the same mass, because water consists of two hydrogen atoms and one oxygen atom, so they must have a greater mass than a single hydrogen atom.
3. *Analysis/Chemistry.* The lithium atom has one electron in the highest energy level: three electrons minus two electrons in first energy level equals one electron in the next energy level. Hydrogen has only one electron, which is therefore also in the highest energy level; in hydrogen's case, this is also the first energy level. (This similarity among all the elements of the first group accounts for specific properties that are common to these elements.)
4. *Application/Chemistry.* Yes, calcium is probably a good conductor of heat and electricity because it is on the left side of the periodic table where the metals are found, and thus has the properties of a metal.

Lesson 4

Substances can be combined as mixtures, which involve physical changes, but not chemical changes. A mixture in which one substance is dissolved in another substance is called a solution.

1. *Application/Chemistry.* a. A soft drink is a solution in which a gas (carbon dioxide) is dissolved in a liquid. b. Salted popcorn is a mixture because you can still see the salt on the popcorn. c. Lemonade is a solution of lemon, sugar, and water. The sugar is dissolved in the lemon and water. d. Sea water is a solution, because salt is dissolved in the water.
2. *Comprehension/Chemistry.* No, because sugar dissolves into molecules in solution, and so must be a nonelectrolyte.
3. *Analysis/Chemistry.* Tap water conducts electricity because it contains ions, such as salts, which are electrolytes.
4. (4) *Evaluation/Chemistry.* If the solution's water were salty before the substance being tested was added, then the solution would always conduct electricity because salt, an electrolyte, was already present. Thus it would be impossible to tell whether or not the additional substance was an electrolyte as well.

Lesson 5

Substances combine with each other through chemical reactions, which involve electron transfer or sharing at the molecular level.

1. *Application/Chemistry.* The substitution must be a chemical reaction because the result is a substance (tooth enamel) that has different properties from the original. The new tooth enamel is harder and more resistant to decay.
2. (5) *Comprehension/Chemistry.* $2O_2$ represents two molecules of oxygen, which is an element. In each of the other choices, more than one element appears in the chemical formula, indicating that it is a compound.
3. *Comprehension/Chemistry.* The reactants in both reactions are gasoline (carbon and hydrogen) and oxygen gas. The products of the first reaction are carbon monoxide and water vapor. The products of the second reaction are carbon dioxide and water vapor.
4. (2) *Analysis/Chemistry.* Under certain conditions, the combustion of gasoline yields CO; under other conditions it yields CO_2. The difference between the two gases is the number of oxygen atoms. Because CO_2 has one more oxygen atom in its formula, it seems likely that increasing the oxygen available during combustion would yield more carbon dioxide, and thus less carbon monoxide.

Lesson 6

Salts and water are the products of a chemical reaction between acids and bases.

1. (1) *Application/Chemistry.* A base is a substance that has a pH greater than 7; to grow lettuce you need soil with a pH greater than 7.
2. (4) *Evaluation/Chemistry.* The correct answer is (4) because acids are electrolytes, and stronger acids conduct electricity better than weaker acids. Choices (1), (2), and (3) are incorrect because adding new substances to the solution will change its properties. Choice (5) is incorrect because it can be extremely dangerous, as well as inaccurate, to test any unknown substance in this way.
3. *Analysis/Chemistry.* The neutralization reaction of sodium bicarbonate and stomach acid yields a sodium-containing salt, which should be avoided by people who have high blood pressure. You can be sure that the sodium from the baking soda is in the salt after the reaction is completed, because water (H_2O) does not contain sodium, and the law of conservation of matter states that the total number of atoms in a chemical reaction does not change; instead, the atoms are rearranged.

Lesson 7

In certain chemical reactions, called exothermic reactions, energy is given off and transferred to the environment as heat, light, or electricity. In other chemical reactions, called endothermic reactions, the substances involved absorb energy from the environment.

1. *Analysis/Chemistry.* Energy is given off by an exothermic reaction, and thus is a product, so it

would be written on the right side of a chemical reaction. Energy is absorbed during an endothermic reaction, and thus it would be written on the left side of the equation with the reactants.

2. *Application/Chemistry.* An ice-cube requires heat to melt. It absorbs this heat from the environment, and so is undergoing an endothermic reaction.
3. *Application/Chemistry.* Because the reaction that breaks up water is one in which a compound (H_2O) is broken up into its component elements, it is endothermic. Thus, energy is necessary for the reaction to take place.
4. *Comprehension/Chemistry.* Entropy has been increased, because the ink is now randomly distributed throughout the water.

Lesson 8

An element can change to another element if it undergoes radioactive decay. In radioactive decay, the nucleus of an unstable—or radioactive—atom, emits protons, neutrons, or electrons. When an atom loses protons, it changes to another element.

1. (5) *Application/Chemistry.* According to the periodic table, plutonium has an atomic number of 94. The lesson states that all elements that have an atomic number greater than 83 are radioactive. The other choices are all elements that have atomic numbers less than 83.
2. *Analysis/Chemistry.* The number of protons in an atom, or its atomic number, determines which element it is. The only element that has two protons is helium.
3. *Analysis/Chemistry.* As the nuclear fuel in a nuclear reactor undergoes radioactive decay, it gradually turns into nonradioactive elements. These new elements will not produce nuclear energy, so they must be replaced with more radioactive fuel.
4. (2) *Evaluation/Chemistry.* Carbon dating may only be done on tissue that was once living. A rock is not alive and never was, so it is impossible to use the technique on a rock.

Chapter 4 Quiz

1. *Application/Chemistry.* Ten protons would yield ten positive charges, 12 neutrons would yield no charges, and 11 electrons would yield 11 negative charges for a net charge of -1.
2. (3) *Analysis/Chemistry.* The vitamin A molecule is broken down during a chemical reaction because of the energy supplied by sunlight. Choices (1) and (4) are incorrect because the milk container is closed, so there is nowhere for the vitamin or the milk to evaporate to. Choices (2) and (5) are incorrect because there is no way to tell whether they are true of every milk solution, not just ones left in sunlight.
3. *Comprehension/Chemistry.* With a pH of 7.0, water would fall on the scale between blood and milk.
4. *Analysis/Chemistry.* An acidic stomach would probably be least helped by another strong acid, because such a substance would not serve to neutralize the acid in any way. The food that has the strongest acid content on the scale is lemon juice.
5. (4) *Evaluation/Chemistry.* A catalyst is a chemical which speeds up the rate at which a chemical reaction proceeds. Thus the fact that papain breaks up fibers more quickly than cooking, a chemical reaction, is evidence that papain is a catalyst.
6. *Application/Chemistry.* A catalyst was most likely added at about three hours, the point when the amount of the reaction products formed increased dramatically, indicating the speed of the reaction increased.
7. (2) *Application/Chemistry.* A neutralization reaction occurs when an acid reacts with a base, indicating calcium carbonate acts as a base in neutralizing acidic lakes.
8. *Analysis/Chemistry.* When you add ice to the soda, you lower the temperature of the solution, which makes the carbon dioxide less soluble. The excess gas escapes from the solution in the form of bubbles or "fizz."
9. (1) *Comprehension/Chemistry.* Electrons whirl around their nuclei in clouds.
10. (3) *Evaluation/Chemistry.* Ordinary waste containers do not contain 18 inches of lead or 32 inches of concrete. There is no basis for choices (1), (4), and (5). Choice (2) is not a problem, as the power plant managers could order more lead and concrete if necessary.

Chapter 5 Physics

Lesson 1

Work is done whenever a force causes an object to move. Power is the rate at which work is done. Energy is the ability to do work.

1. *Application/Physics.* a. A growing plant is moving, so it exhibits kinetic energy. b. A rotating galaxy is moving, so it exhibits kinetic energy. c. A dish on a table is not moving, so it exhibits potential energy.
2. *Application/Physics.* Energy from the sun (solar energy) is changed to electric energy as it strikes the cell. This electric energy then becomes mechanical energy as it powers the calculator.
3. *Comprehension/Physics.* Power is the rate at which work is being done. The person running up stairs is doing the work of moving up the stairs more quickly than the walking person, so the person running is using the most power.
4. (3) *Evaluation/Physics.* Choice (1) is implied by the definition of energy. Choices (2) and (4) are ways in which energy is converted from one form to another. However, choice (3) is contradicted by the law of conservation of energy—that energy can neither be created nor destroyed.

Lesson 2

The forces that affect motion are those that cause motion and those that tend to resist motion. Gravity, friction, and inertia are forces that commonly act upon us.

1. *Application/Physics.* You are applying friction, in this case between the brake and the wheel. The friction between the two opposes the forward motion of the car, eventually bringing the car to a stop.
2. *Analysis/Physics.* According to Newton's first law, as a car comes to a sudden stop, people inside the

car continue to move forward due to their own inertia. Their seat belts act as an outside force to halt their forward motion, preventing serious injury in most cases.

3. *Analysis/Physics.* According to Newton's second law, it takes more force to move a large object than to move a small one. More force requires more energy, which is provided by fuel. Thus a smaller car can be powered by less fuel than a larger car, which also makes it more economical.

Lesson 3

Heat and mechanical energy are both forms of energy. Heat can be converted to mechanical energy in order to do work, and mechanical energy can be changed into heat through friction.

1. *Comprehension/Physics.* Heat that is converted into mechanical energy is able to do work.
2. *Analysis/Physics.* The mechanical energy of the slide produces friction between the hands and the rope and is thus converted to heat, causing the hands to be burned. At the same time, friction removes cells from the hands, causing part of the burn effect through abrasion of the hands.
3. *Application/Physics.* According to the second law of thermodynamics, heat will flow from the hotter object to the cooler object. In this case, heat flows from the hot tea to the ice cube.
4. (2) *Analysis/Physics.* In an ordinary automobile, friction between the moving parts decreases the amount of energy available as mechanical energy to move the car. Oil allows the parts to slip past each other more easily, thus increasing the amount of usable mechanical energy. Choices (1) and (3) are incorrect because lowering the temperature of the gases or engine would decrease the amount of heat energy available and thus decrease the available mechanical energy. Choice (4) is incorrect because it addresses the problem of friction between the car and the outside environment, not the friction between parts of the car itself.

Lesson 4

All waves share the properties of having an amplitude, a length, and a frequency. In addition, all have a characteristic shape which includes a crest and a trough.

1. (3) *Application/Physics.* Because the moon has no atmosphere, sound waves do not have a medium through which to travel, so the radio would not be useful. Because light does not require a medium through which to travel, a flashlight (choice [1]) and a camera (choice [2]) would both be useful. Thermal underwear (choice [4]) would also be useful in retaining the body's heat energy.
2. *Analysis/Physics.* Sound waves travel faster through solids than through liquids or gases. Thus the few feet of air between the TV set and the wall would slow down the movement of sound into the next apartment, making such an arrangement less likely to disturb a neighbor.
3. (4) *Evaluation/Physics.* The relative speeds at which light waves and sound waves travel is not discussed in the passage. Many of the waves mentioned, such as sound waves, are invisible, eliminating choice (1). The fact that we can see light or hear sound from across the room implies choice (2), eliminating it as well. The fact that light from the sun travels to Earth implies and thus eliminates choice (3). Choice (5) is eliminated because a long wavelength yields a low frequency which in turn is low in energy.
4. *Application/Physics.* A brighter light is to the greater amplitude of a light wave as a louder sound is to the greater amplitude of a sound wave. Both imply an increase in the intensity of the wave.

Lesson 5

Light has the properties of reflection and refraction. When light is reflected, it bounces off an object. Light that is refracted is being bent. Light is also able to travel at speeds as high as 186,000 miles per second.

1. *Comprehension/Physics.* The pencil is being bent as it enters the water, and thus is illustrating the property of refraction.
2. *Application/Physics.* No. Because of the refraction of light, the ring is not exactly where it appears to be. You would have to determine how the light was being bent, and then dive at an angle to the object's apparent location.
3. (1) *Comprehension/Physics.* It is the different frequency of different colors of light that causes each color to refract at a different angle.
4. (2) *Evaluation/Physics.* Light is a form of electromagnetic radiation, so the properties of light are probably the properties of all electromagnetic radiation. Choice (1) is incorrect because the sun emits light, a form of electromagnetic radiation. Choice (3) is incorrect because it is true of light. Choices (4) and (5) are not mentioned in the passage, and are in fact incorrect.

Lesson 6

Electricity is a basic property of all matter because matter is made of atoms, and atoms contain electrically charged particles.

1. *Comprehension/Physics.* Static electricity is an uncontrolled transfer of electric charges. Current electricity is the controlled flow of electric charges.
2. *Analysis/Physics.* Electric charges build up on your shoes as they rub against the carpet. The shock results as a sudden transfer of electrons occurs between your shoes and the doorknob.
3. (2) *Analysis/Physics.* Because energy can neither be created nor destroyed, eliminating choice (1) the energy must have been lost to the environment in a different form. Choice (3) is incorrect because 100 percent of the electrical energy has been used by the coffee pot in some way, implying it was converted *from* potential energy, although only 90 percent of it was converted *to* heat energy.

Lesson 7

Magnetism is related to electricity through electromagnetism. Electromagnetism involves the use of magnets to generate an electric current, or the use of an electric current to create a magnetic field.

1. (4) *Application/Physics.* Because the force lines indicate repulsion, like poles of the two magnets must be near each other. Choice **(1)** is incorrect because magnets must be close to each other in order to exert a force on each other. Choices **(2)** and **(3)** are incorrect because un-

like poles attract each other. Choice (5) is incorrect because the force lines from both magnets are of equal size, so the magnets must be of equal strength.

2. *Evaluation/Physics.* Since opposite poles attract, the point on a compass marked north must actually be the opposite of Earth's north pole. It is a south pole.
3. *Analysis/Physics.* Because copper is used to generate electricity, it must allow current to flow freely, and is thus a conductor. (An example of an insulator would be rubber or glass.)

Lesson 8

Matter is converted to energy through nuclear fission or nuclear fusion. Nuclear fission is the splitting apart of an atom. Nuclear fusion is the joining together of two atomic nuclei.

1. *Comprehension/Physics.* Both fission reactions and fusion reactions take place in the nuclei of atoms, and both release energy that is contained in the atomic nucleus. Fission reactions involve the splitting of atomic nuclei, while fusion reactions involve the joining of atomic nuclei. Fission reactions are created by human beings, and they can be controlled. Fusion reactions occur spontaneously on the surface of the sun and other stars, but are difficult for human beings to create. So far, they cannot be controlled.
2. (1) *Application/Physics.* In a chain reaction, one neutron splits an atom, which releases more neutrons, which go on to split other atoms, which release more neutrons, and so on. Choices (2) and (3) are incorrect because the situations do not involve a multiple increase such as 1 to 2 to 4 to 8, etc. Choice (4) describes a reduction, not an increase.
3. *Analysis/Physics.* Nuclear power could threaten the environment if radiation leaked from a reactor, if a reactor overheated, or if the reaction proceeded out of control.

Chapter 5 Quiz

1. *Analysis/Physics.* The figure shows weight (1) remaining at rest vertically after knocking weight (2) to the right. This violates Newton's third law: an equal and opposite reaction would cause weight (1) to swing back toward the left.
2. (1) *Analysis/Physics.* Refraction (choice [3]) would cause the surface of the water to shimmer. However, reflection of the sunlight off the water into your eyes is what would cause problems in seeing where you were going. Choices (2), (4), and (5) would not cause a glare, and so would not interfere with your ability to see where you are going.
3. *Application/Physics.* Choices (b), (c), (e), and (f) all involve movement of an object in some way, and so are work according to the physics definition of work.
4. (3) *Evaluation/Physics.* While the other choices are true, only choice (3) shows electricity generating a magnetic field, which is an example of the interrelationship of the two.
5. *Application/Physics.* Baking is an example of heat flowing from a hotter object (the oven) to a cooler one (the cookies), so it is an application of the second law of thermodynamics.
6. (5) *Analysis/Physics.* Statements A and D are based on someone's judgment, while statements B and C can be tested in a controlled environment.
7. (4) *Evaluation/Physics.* Statements B and C can be experimentally proven, while neither statements A nor D can be subjected to a conclusive test.
8. *Comprehensive/Physics.* Wave A has the greater amplitude, or high points and low points, and wave B has the greater wavelength, or distance between one crest of the wave and the next.
9. *Application/Physics.* Wave A has the greater energy, becuase waves of shorter wavelength have greater frequency and greater energy.
10. *Comprehension/Physics.* The statement is TRUE. Nuclear fission reactions are initiated by human beings in scientific laboratories or in nuclear reactors. Fusion reactions take place spontaneously on the surface of the sun and other stars.

Chapter 6 Interrelationships Among the Sciences

Lesson 1

Biology, earth science, chemistry, and physics have the scientific method in common. In addition, the various branches share topics; for example, biochemistry is of interest to both biologists and chemists. Science also influences other fields of study and is in turn influenced by them.

1. *Evaluation/Earth Science.* Twenty-thousand years ago, the mountaintop was under water. The fish lived in the water a few thousand years before that, along with many others of its species.
2. *Analysis/Chemistry.* Through the use of carbon-14 dating, the second scientist proved that the fish fossil was much older than 5000 years old.
3. *Application/Earth Science.* The first scientist was an earth scientist, someone who studies the forces which shape Earth. The second scientist was probably a chemist or a physicist; both fields involve the study of radioactivity. The third scientist was a biologist, someone who studies life on Earth.
4. *Evaluation/Earth Science.* If the mountain is growing at the rate of six inches per year, many thousands of years ago it was much lower. The fish fossil indirectly supports this theory by showing that at the time that the fish was alive, the mountain was under water, indicating that it has risen since then (assuming the level of the water has remained constant for the past 25,000 years).

Lesson 2

Knowledge of chemistry and physics combines to send a rocket ship into outer space. Astronomers value space exploration for the firsthand knowledge it provides about the solar system. Biologists are concerned with keeping people alive in the environment of outer space.

1. (3) *Application/Biology.* Biology deals with living organisms and the effects of the environment on life.
2. *Evaluation/Biology.* Yes, because the lesson describes an experiment

on Earth in which plants were used to maintain a liveable environment over a period of five months. It is reasonable to assume this experiment could be duplicated on a long voyage.

3. *Analysis/Biology.* Some other considerations in planning a garden for a spaceship are the amount of space the plant takes up, its taste, and how easily it is stored.

Lesson 3

Scientists can predict air pollution and its possible effects. Both scientists and engineers use scientific principles to design ways to reduce air pollution.

1. (1) *Comprehension/Earth Science.* Natural dusts account for 63,000,000 tons of pollution per year, and thus are the largest source.
2. *Analysis/Earth Science.* Transportation accounts for 1.2 million tons of emissions per year, while the burning of forest litter accounts for 11 million tons. Thus stopping the controlled burning of forest litter would reduce much more air pollution than switching to a nonpolluting form of transportation would.
3. *Comprehension/Earth Science.* A sudden buildup of pollutants suggests there was an atmospheric inversion above London at that time.
4. *Application/Earth Science.* The particles from the volcano were probably carried by wind around the world to New England, where they caused cold temperatures by lowering the amount of sunlight that reached Earth at that point.

Chapter 6 Quiz

1. (4) *Application/Physics.* The study of motion and forces in physics includes speeds and pressures of fluids.
2. *Comprehension/Physics.* The action of the runners against the starting blocks and the ground and the resulting reaction of Earth is described by Newton's third law of motion, an important law of physics. A physicist would be interested in the description as a manifestation of Newton's third law.
3. *Comprehension/Biology.* Digestion is the breakdown of foods into nutrients that the body can use.
4. *Application/Biology.* a. Chewing food requires movement, so it is an example of mechanical digestion. b. Digesting proteins into amino acids is the breaking apart of a food molecule, or chemical digestion. c. The movements of the stomach are mechanical digestion. d. The action of bile on fats is chemical digestion.
5. *Evaluation/Chemistry.* Yes, it is evidence, because saliva digests complex carbohydrates into sugars. Because the cracker only begins to taste sweet after it has been in your mouth for a time, the saliva has probably digested complex carbohydrates into simple sugars during this time.
6. (2) *Analysis/Biology.* Although all the choices involve life processes, a biologist would probably be most interested in the fact that the final products of digestion are in a form the body can use. Choices (1) and (4) would be of most interest to a chemist. Choices (3) and (5) are of most interest to a physicist.
7. *Analysis/Biology.* The fact that metabolism is a body process makes it of interest to a biologist. The fact that it is a process that changes one substance to another with different properties means that it is a chemical reaction, and thus is of interest to chemists as well.
8. (4) *Application/Earth Science.* These conditions are the only ones of the choices that yield a wind chill of below $-35°$ F, the point at which the National Weather Service issues a special advisory.
9. (3) *Analysis/Earth Science.* The wind-chill factor takes into account both winds and temperature. For example, according to the figure, a calm day at 10° F has a higher wind-chill factor than a 10° F day with higher winds. Thus a calm day feels warmer than a day of the same temperature that is also very windy.
10. *Evaluation/Chemistry.* Human flesh is not made up of the same chemicals as plastic and water, although it does contain some water. However, other chemicals such as protein and fats may affect the rate at which human flesh freezes, meaning that it freezes at a faster rate or a slower rate than the water in plastic containers did in the experiments. Thus, the error lies in applying the results of tests on one chemical (water) to other chemicals (those in the human body).

UNIT II Test

1. *Comprehension/Physics.* The substance remains carbon dioxide even though it has changed from a solid to a gas. Thus, this is a physical change.
2. (2) *Application/Physics.* Only burning wood involves the formation of new substances, in this case, ashes and gas.
3. *Comprehension/Earth Science.* A glacier forms hills by pushing before it and to the sides mounds of earth and rubble. It forms valleys with the rocks and pebbles attached to its bottom and sides, which aid it in scraping away soil.
4. *Application/Earth Science.* B shows a U-shaped valley, typical of those formed by glaciers.
5. *Comprehension/Earth Science.* The rocks were probably brought from a distant region by an ice sheet that carried them and then left them behind when it melted.
6. (5) *Analysis/Earth Science.* Earth's inward force, or gravity, causes the downward slide of glaciers. Gravity also pulls snow toward the center of Earth, aiding its transformation into ice.
7. *Analysis/Earth Science.* The level of the oceans would rise with the water from the melting glaciers. Land near the oceans would be flooded.
8. (2) *Application/Biology.* A 2:4 chance, or $\frac{1}{2}$, exists for Tt, which would produce a tall offspring, because T is a dominant gene.
9. (3) *Evaluation/Chemistry.* Statements A and C are based on opinion, while B and D could be proven by chemical testing.
10. (2) *Comprehension/Biology.* The highest weight given for this height is 123 pounds.
11. *Application/Biology.* No, because this weight falls within the range of desirable weights for this height and age.
12. (2) *Analysis/Biology.* Bone structure is not a factor in either table.
13. *Evaluation/Biology.* Yes; because

women have the same values as the men in the GRC table, they could weigh the same as a man their height and age, even though they have lighter bones and muscles. This means they could be fatter than a man the same age, and still fall within the table's "desirable" weight range.

14. *Evaluation/Biology.* The Gerontology Research Center table would tend to support the statement because it suggests that young adults should be lighter than older adults. Thus a 5-foot-9-inch person who weighed 200 pounds at age 25 would have an undesirable weight on this model, whereas a 5-foot-9-inch person who weighed 200 pounds at age 65 would be within the desirable weight range for that age.
15. (5) *Comprehension/Physics.* According to the passage, air expands and takes up more space, cooling as it does so.
16. (1) *Analysis/Physics.* At sea level, air pressure is greater, so more energy is needed to boil water, thus allowing the molecules to escape the liquid. This means boiling water is hotter at sea level, so cooking time would be shorter.
17. *Analysis/Physics.* Electrical charges exist separately and can be isolated from each other. But if a magnet is cut in half, the result is two complete magnets, each with a north pole and a south pole. This continues through all further subdivisions.
18. (1) *Analysis/Biology.* Initial testing outside of live animals must assume that conditions and results will be similar.
19. (2) *Evaluation/Biology.* It is logical to expect that the chemicals will bind to the fiber in about the same amount as they did in the test tube, at least until proven otherwise.
20. *Evaluation/Biology.* Yes, because the 8 percent some bound is a small amount. Ninety-two percent of the chemical remains unbound and potentially harmful. On the other hand, whole sorghum flour bound 50 percent of the chemicals, and thus would provide a significant protection against carcinogens.
21. (5) *Analysis/Biology. Protoavis* seems to be a transitional form between dinosaurs and birds, sharing features of each. Since birds evolved much later than dinosaurs, the fossils support the theory that birds evolved directly from dinosaurs and not from a more distant common ancestor.
22. (3) *Analysis/Biology.* Sharks are ocean-dwelling animals, therefore, the presence of shark fossils indicates that an ocean existed in that area at one time.
23. (1) *Analysis/Earth Science.* Shock waves travel at different speeds through materials of different density. Thus, different rock formations must vary in density for seismographs to be useful in locating petroleum deposits.
24. *Application/Chemistry.* No, because a compound that has a high heat of formation is very stable.
25. *Application/Chemistry.* NaCl is more stable than HgO because of NaCl's higher heat of formation.
26. *Comprehension/Chemistry.* The compound is probably very unstable because it has a negative heat of formation.
27. (4) *Application/Chemistry.* A compound that has a negative heat of formation absorbs energy, so it is formed by an endothermic reaction. A compound that has a positive heat of formation releases energy, so it is an exothermic reaction.
28. (3) *Analysis/Chemistry.* The reaction occurs at high temperature, so energy is added, meaning the reaction has a negative heat of formation. Such a reaction yields an unstable compound.
29. (1) *Evaluation/Biology.* Bacteria break down all dead matter, releasing it back into the environment for use, so are most important in continuing the cycling of matter. Phytoplankton are also important in that they produce food, but they depend on nutrients the bacteria supply.
30. *Application/Biology.* A human would be a consumer, because humans eat other living things for food, as do the fish and mussels.
31. *Analysis/Biology.* There are several effects: The gulls will be forced to eat more shrimp to make up for the lack of mussels. Overconsumption of shrimp may cause a decline in fish and gulls. Both types of plankton may temporarily increase, because less shrimp and mussels exist to consume them.
32. (4) *Application/Physics.* According to the figure, a magnetic force field is produced when an electric current is flowing. Only the toaster and flashlight use electricity.
33. *Analysis/Physics.* The electric current in (B) is flowing in the opposite direction from the current in (C). From this observation, a hypothesis may be formed that the lines of force of a magnetic field are dependent on the direction in which the electric current is flowing.

PRACTICE

Practice Items

1. (3) *Application/Biology.* A sperm is a gamete. Each gamete contains one-half the number of chromosomes as its parent cell. If the parent cell has 46 chromosomes, the gamete must have 23 chromosomes.
2. (4) *Application/Biology.* An egg is a gamete and contains one-half the number of chromosomes as its parent cell. If the egg had four chromosomes, the parent cell must have eight chromosomes.
3. (4) *Application/Biology.* Because both frog egg and sperm are gametes, each contains one-half the number of chromosomes as its parent cell. Because the gametes have joined, the chromosome count in the developing young equals that of either of its parents.
4. (2) *Analysis/Biology.* Resemblance between offspring and parent is not due to the number of chromosomes. A human child receives 23 chromosomes from each parent, regardless of which parent he or she comes to resemble. All the other choices deal with changes in the number of chromosomes.
5. (1) *Evaluation/Biology.* The passage states that, before it divides the parent cell increases in size. The daughter cells, each of which contains half of the enlarged parent's cellular material, must be smaller than the enlarged parent cell. The other choices are not addressed in the passage. Of them, however, Choices (2) (3), and (4) are false. Mutations, which are discussed in Choice (5), can arise from several causes.
6. (3) *Comprehension/Biology.* The movement of animals into better feeding grounds may be due to a number of outside factors. For example, drought or an overpopulation of animals could drive animals to search for a better food supply.
7. (4) *Comprehension/Biology.* A body's alertness at an unusual hour is one manifestation of a disrupted biological clock. Choices (1), (2), and (3) illustrate the proper working of the biological clock. Choice (5) describes a situation that is unrelated to the workings of biological clocks.
8. (3) *Comprehension/Biology.* As shown in the figure, the uncorrected vision results in the focusing of the image behind the retina. When a corrective convex lens is used, the image is focused on the retina.
9. (3) *Analysis/Biology.* Because flowers do not carry out photosynthesis, the sugar in nectar must be produced in a part of the plant that contains chlorophyll and then transported to the flower. Only phloem tissue carries the products of photosynthesis.
10. (2) *Analysis/Biology.* The phloem tissue, the kind destroyed by gnawing, carries food from the leaves to other parts of the plant. Only choice (2) correctly describes destruction of that process. Choice (3) describes the process that involves xylem tissue. Choices (1) and (4) describe processes incorrectly. Choice (5) names a process that would not be directly affected by gnawing.
11. (3) *Analysis/Biology.* Because identical twins develop from the same union of sperm and egg, the twins have identical chromosomes, including the combination that determines gender. (The gender of the other family members is irrelevant to the twins' gender (Choice [1]). Fraternal twins need not be of the same gender (Choice [2]), nor need they resemble one another or each parent closely (Choices [4] and [5]).
12. (5) *Comprehension/Biology.* As specified in the passage, the classification system that was developed by Linnaeus names the genus and the species of an organism.
13. (1) *Application/Biology.* If two organisms belong to the same species, they can breed and produce offspring. None of the other choices relate to the species of the organisms.
14. (5) *Application/Biology.* If the alleged parents do not belong to the same species, they cannot interbreed.
15. (4) *Application/Biology.* At the given rate of reproduction, there would be 20 staphylococci in the soup by 12:20 P.M., 40 by 12:40 P.M., and 80 by 1:00 P.M.
16. (3) *Comprehension/Biology.* The passage suggests the opposite of Choice (3)—namely, that even renewable resources can be used more rapidly than they can be replaced.
17. (1) *Evaluation/Biology.* The passage suggests that the rate of population growth is directly related to the rate at which resources are consumed. Choice (2) is contradicted by the passage, which implies that small populations do not overtax natural resources. (3) and (4) are not addressed in the passage. Fishing, which is discussed in Choice (5), is regarded as dangerous only when overfishing takes place.
18. (3) *Analysis/Biology.* According to the passage, once the plants have been removed, the actions of wind and water lead to soil loss. Drought and the actions of humans are the major causes of plant loss. The other choices describe situations that would account for plant damage and resulting soil loss in some countries, but would not be true for all.
19. (4) *Comprehension/Biology.* According to the passage, only one species makes up a given population. Small mammals, which are

discussed in Choice (4), would comprise several species. All the other choices involve only one species.

20. (2) *Application/Biology.* With the increase of vegetation in the area, the pond eventually will be filled in, and the local community will change. The factor of pollution, discussed in Choices (4) and (5), is not considered in the passage.
21. (4) *Analysis/Biology.* If the process of succession continues unhindered, the pond would be replaced by a bog, then a meadow, and then a forest—the climax community. None of the choices describe the climax community itself. Of the choices that are given, Choice (4) describes a meadow, the most developed stage before that of a forest.
22. (4) *Evaluation/Biology.* Of the environments that are listed, only that of hardwood forests describes a typical climax community. Choice (5) describes a habitat that was built by humans, not created by nature.
23. (3) *Application/Biology.* According to the figure, sweetness is sensed on the tip of the tongue. Other tastes are sensed on the sides and the back of the tongue. No tastes are sensed on the center of the tongue.
24. (1) *Analysis/Biology.* Eating seeds instead of planting them will relieve the food shortage only temporarily. In the future, the problem will become worse, due to the lack of both crops and seeds to grow crops.
25. (3) *Analysis/Biology.* According to the equation, the end products of respiration are the products that are used in the process of photosynthesis.
26. (2) *Evaluation/Biology.* The passage describes attempts to reduce noise and air pollution. Choices (1), (3), and (5) deal with the protection of the cement wall, not with a pollution problem. Choice (4) deals with protecting soil, not the people living nearby.
27. (2) *Analysis/Biology.* No involvement of the brain is suggested in the figure, but the involvement of the spinal cord and the nerves (bundles of neurons) is clear. The source and the speed of simple reflexes, which are discussed in Choices (4) and (5), are not considered in the figure.
28. (4) *Application/Biology.* Shallow, branching roots would enable a plant to absorb water rapidly during infrequent desert rains. The other mutations would be disadvantages—a grasshopper that is unable to fly to find food and avoid prey, a mouse that stands out sharply from its surroundings, a giraffe that is unable to reach for leaves in trees, and a lion that is unable to brush away harmful insects.
29. (3) *Evaluation/Biology.* A harmful mutation makes an organism less able to compete and thus more likely to die. Choices (1), (4), and (5) are possible but are far less likely than Choice (3). Choice (2) is very unlikely, since mutations are random occurrences.
30. (5) *Comprehension/Biology.* Using the method of determining probability that is illustrated in the figure, these are the probable outcomes of the choices given.
    Choice (1): bb × bb = bb + bb + bb + bb (all white)
    Choice (2): bb × Bb = bb + bb + Bb + Bb (half white, half blue-gray)
    Choice (3): BB × BB = BB + BB + BB + BB (all black)
    Choice (4): BB × Bb = BB + BB + Bb + Bb (half black, half blue-gray)
    Choice (5): BB × bb = Bb + Bb + Bb + Bb (all blue-gray)
31. (2) *Evaluation/Biology.* The chief characteristic of this relationship is that both organisms benefit from the symbiosis and neither organism harms the other. This characteristic eliminates Choice (3). The passage suggests nothing about the continuing nature of symbiosis (Choice [1]), the kingdom of the organisms that are involved (Choice [4]), or the complexity of the organisms that are involved (Choice [5]).
32. (5) *Evaluation/Biology.* The lichen's algae produce food, which the fungi part needs to exist. Without that food, the fungi will die. This fact eliminates Choices (1) and (4). Choices (2) and (3) refer to information about lichen that is not discussed in the passage.
33. (5) *Comprehension/Biology.* According to the figure, the other choices name structures that are found in the cells of both plants and animals.
34. (4) *Comprehension/Earth Science.* According to the time line, mammals appeared during the first period of the Mesozoic Era, and developed further into placental mammals in the third period. Wooly mammoths, which are discussed in Choice (2), could have appeared during the Mesozoic Era, but they did not become extinct during that era. All the other events that are listed relate to the Cenozoic Era.
35. (2) *Analysis/Earth Science.* Dinosaurs died out about 70 million years ago, during the Cretaceous Period. Discussion of the past million years of Earth's history would not include dinosaurs.
36. (1) *Application/Earth Science.* Because horses appeared less than 100 million years ago, it is impossible to have a horse bone that is 150 million years old.
37. (3) *Evaluation/Earth Science.* According to the time line, the Andes were formed during the Tertiary Period. Thus, they are younger than the Rockies, which were formed during the more distant Jurassic Period.
38. (4) *Analysis/Earth Science.* Because magnesium is just one of twelve listed elements that together make up 0.5% of ocean water, it is impossible to tell from the chart exactly how much magnesium would be present in any given sample of ocean water.
39. (2) *Application/Earth Science.* The hardness of the porcelain forms a suitable medium for testing the colored powder that has been left by a mineral.
40. (4) *Application/Earth Science.* An object that is less dense than water would float, but an object that is more dense than water would sink. Because the mineral samples are the same size, they must differ in density.
41. (1) *Application/Earth Science.* It is the reflection of light from the surface of the mineral that causes pyrite to resemble gold.
42. (5) *Application/Earth Science.* Mica is characterized by its tendency to break in layers—that is, along definite lines, or cleavage.
43. (2) *Application/Earth Science.*

The Mohs Scale ranks minerals that have high resistance to being scratched—an evidence of the hardness of the mineral—above minerals that have lower resistance.

44. (4) *Comprehension/Earth Science.* Although Choice (1) is a plausible answer, the passage clearly emphasizes the need for quick burial before decay or ravaging by other organisms takes place.

45. (1) *Analysis/Earth Science.* Igneous rock formation involves temperatures that are high enough to melt rock; metamorphic rocks are formed under intense heat and pressure. The temperature and the pressure within Earth must be much greater than anything experienced on the surface of Earth. Choices (4) and (5) address characteristics that are not discussed in the passage.

46. (4) *Analysis/Chemistry.* The decrease in pressure that is caused by opening the bottle plus the increase in temperature that is caused by the warmth of the sunny window will cause the soda to be able to hold less gas than before. Choice (1) would be true only if the bottle remained closed, for gas under pressure does expand when heated.

47. (1) *Comprehension/Chemistry.* The passage names acids and bases as examples of compounds.

48. (3) *Application/Chemistry.* The passage states that HCl + NaOH = NaCl. It would be reasonable to assume that HCl + KOH = KCl. Water would be the other product of both reactions.

49. (5) *Analysis/Chemistry.* The passage states that a base will turn litmus paper blue. Bases can be powerful corrosives just as acids can be; for example, lye is mentioned as an example of a base.

50. (2) *Evaluation/Chemistry.* Some 10 million molecules of hydrochloric acid (HCl) are left. Excess, concentrated amounts of HCl will eat away at body tissues, making it unsafe to drink.

51. (4) *Comprehension/Chemistry.* Because the specific heat of ice is half that of water, it takes only half as much heat to raise the temperature of 1 gram of ice 1°C. Thus, ice will increase in temperature twice as fast as water.

52. (2) *Analysis/Chemistry.* The steam decreases in temperature by 235°C. Because the specific heat of steam is 0.5, the heat lost per gram of steam is equal to 235 × 0.5 = 117.5 calories.

53. (5) *Analysis/Chemistry.* Heat gained is equal to mass × specific heat × change in temperature. Because both iron blocks received (and gained) the same amount of heat and both have a specific heat of 0.107, one can write: heat gained by A = heat gained by B; mass A × 0.107 × 50° = mass B × 0.107 × 150°. Solving for the ratio of mass A to mass B yields, mass A = $\frac{150}{50}$ mass B, or mass A is three times as great as mass B.

54. (2) *Evaluation/Chemistry.* Because a gas has neither definite shape nor definite volume, perfume vapors could transfer easily from a small bottle to a large room. Choice (4) is incorrect because the gas would expand immediately to fill the 50-liter tank upon being released from the balloon.

55. (3) *Analysis/Chemistry.* It is clear from the equation that the bonds A-X and B-Y must be broken in order to form the bonds A-Y and B-X. If energy is released by the reaction, then the energy that is required to break bonds must be less than the energy that is released in forming bonds.

56. (3) *Evaluation/Physics.* Before work is begun, the objects that will be involved in that work are at rest. They have potential energy. When they are put into motion, they expend kinetic energy. As a result, work is performed.

57. (3) *Analysis/Physics.* If the weight (which is proportional to the mass) increases in the ratio of $\frac{3}{2}$ ($\frac{12}{8}$, expressed in lowest terms), then the acceleration must decrease by a factor of $\frac{2}{3}$ if the force is to remain constant. Thus the bowling ball will gain speed $\frac{2}{3}$ as fast as usual.

58. (4) *Comprehension/Physics.* The work that is done on the box by the worker is equal to 200 newtons × 6 meters = 1200 newton-meters. The work that is done by the crane, which must lift the box straight up with a force that is equal to the weight of the box, is equal to 1000 newtons × 2 meters = 2000 newton-meters. The total work that is done on the box, then, is 1200 newton-meters + 2000 newton-meters = 3200 newton-meters.

59. (5) *Comprehension/Physics.* According to the passage, the weight of a floating object is equal to the weight of the fluid that it displaces.

60. (1) *Application/Physics.* According to the passage, an object floats when its weight is equal to the weight of the displaced fluid. It sinks only until it displaces enough fluid to equal its own weight. In this situation, it is logical to assume that the penny's volume is too small to displace water equal to its weight; thus, the penny sinks to the bottom.

61. (5) *Analysis/Physics.* Water provides greater buoyancy than does air. Absolute weight is not as critical in the water as it is on land because of that buoyancy. Thus the stress produced on the joints during exercise is less in water than in air.

62. (1) *Evaluation/Physics.* A huge, hollow object will displace the most water for its weight. A streamlined design is important when the boat is moving, but is not important to its ability to float.

63. (2) *Application/Physics.* The man's distance from home is not dependent upon how far he walked (distance, Choice [1]) but upon how far his final position is from his starting position (displacement).

64. (5) *Application/Physics.* The velocity of the car always equals the velocity of the truck, but the truck has a greater mass. Therefore, the momentum of the car at any given time will differ from the momentum of the truck.

65. (2) *Application/Physics.* Both friends finished 1 mile from their starting point. Friend 1 covered more distance because of the detour. Friend 1 accelerated, but Friend 2 maintained a constant velocity.

66. (4) *Application/Physics.* Because the velocity of the puck remained constant—2.5 meters per second—and the puck neither started nor stopped, acceleration was zero (0).

Practice Test

1. (3) *Application/Physics.* The pull of gravity on the astronauts decreases as they move farther from Earth (the object that is pulling on them). As the force due to gravity decreases, so does the astronauts' weight. Choices (1), (2), (4), and (5) are false.
2. (4) *Analysis/Earth Science.* Living organisms are found in the layer that contains mountains, oceans, and so on.
3. (1) *Application/Biology.* This is a description of active immunity—acquired naturally. The question implies that the girl is immune (which rules out Choice [5]) and that the immunity is natural because she was exposed when her brother had chicken pox (which rules out Choices [2], [3], and [4]).
4. (3) *Application/Biology.* This is an example of passive immunity—naturally acquired. The infant must be receiving antibodies from its mother's milk.
5. (2) *Application/Biology.* The vaccination will cause the child to produce his or her own antibodies—a demonstration of active immunity that has been induced artificially.
6. (4) *Application/Biology.* The passage implies that the patient's condition requires a very fast immune response. The patient would acquire immunity most rapidly if ready-made antibodies were injected to produce artificially a passive immunity. Choices (1) and (2) would waste time while the patient's body built up its own antibodies, and Choice (3) is unlikely, given the description of the virus. Choice (5) would result in the patient's death.
7. (5) *Application/Biology.* The passage describes a situation in which the immune system has failed to function. Choices (1), (2), (3), and (4) deal with types of immunity and could not account for such total failure of the body's ability to protect itself. A person who has AIDS is susceptible to all diseases.
8. (5) *Analysis/Chemistry.* The equation shows that oxygen is consumed in the process of combustion.
9. (4) *Comprehension/Biology.* Plants receive the energy to make their food from the sun, and all other organisms receive their energy from eating plants (or from eating organisms that eat plants). Therefore, at the most basic level, all living things receive their energy from the sun.
10. (5) *Comprehension/Earth Science.* From the graph it can be seen that the lowest temperatures are present in the thermosphere.
11. (3) *Analysis/Earth Science.* The increase of the ozone layer would tend to keep more heat rays between Earth and the stratosphere. Therefore, the temperature of the troposphere would rise.
12. (2) *Evaluation/Earth Science.* There are so few molecules in the vacuum of space that temperature has no real meaning; even though the molecules have energy, there are not enough molecules to measure. Choices (1), (3), and (5) do not affect the temperature in outer space. Choice (4) is false.
13. (2) *Comprehension/Biology.* The passage states that the nervous system, which includes the spinal cord, is formed from ectoderm cells.
14. (2) *Application/Biology.* Carbohydrates will be used to provide the runner with energy throughout the race, and much water will be lost through perspiration. Fats also will be consumed as the runner calls upon reserves of energy to complete the race, but these molecules will decrease to a much lesser extent than will those of the vital carbohydrates. Minerals and proteins will be the least affected.
15. (3) *Application/Physics.* Dividing distance by time (as described in the second paragraph) gives an average speed of 60 mph (1800 miles divided by 30 hours). The driver, therefore, must have exceeded the 55 mph speed limit at some point in order to complete the trip. Choices (1), (4), and (5) could be true because, although the driver's overall average speed doesn't change, the speed at any single point during the trip could vary. Also, the driver's average speed for different portions of the trip could be different than the overall average speed, so Choice (2) is possible.
16. (2) *Comprehension/Physics.* The passage explains that acceleration includes any change in direction from a straight line. Since Earth orbits in a circle, never in a straight path, it must be accelerating constantly.
17. (3) *Analysis/Physics.* A moving object will continue the same motion unless an outside force causes it to change. In space, there is no air friction, gravity, or other force to stop the ball.
18. (4) *Evaluation/Physics.* A passenger in a car is moving with the car. When the motion of the car is stopped, the passenger will continue to move forward. No other choice explains why the passenger will go forward.
19. (2) *Application/Chemistry.* The passage explains that rusting occurs when iron combines with oxygen and that moisture speeds up the process. Protection against rust, therefore, would involve keeping the object from oxygen (sealing in a vacuum) and from water (dry).
20. (5) *Application/Biology.* If both parents have sickle-cell anemia, then each must have a pair of the disease-causing genes. The disease-causing gene will be the only gene that they can pass on to their children.
21. (1) *Comprehension/Biology.* Saliva in the mouth begins the chemical breakdown of food.
22. (4) *Application/Biology.* Water is

reabsorbed from undigested material in the large intestine. A person who has diarrhea is not reclaiming enough water. Therefore, the large intestine must not be working properly.

23. (3) *Analysis/Biology.* The passage states that one of the functions of the small intestine is absorption of the digested molecules into the bloodstream. The greater the surface area, the more absorption can take place as food moves through. Choices (1) and (2) are not functions of the small intestine. Choices (4) and (5) describe situations that are not helpful for digestion.
24. (4) *Comprehension/Biology.* Of the choices listed, only Choice (4) names two organs, neither of which actually breaks down food: the anus also does not digest, but it is paired with the stomach, which does digest food. Although water is absorbed in the large intestine, absorption is not part of the process of breaking down food.
25. (1) *Evaluation/Biology.* Antibiotics will kill the bacteria in the large intestine as well as those that cause strep throat. As a result, the patient may become deficient in the vitamins that those bacteria produce.
26. (2) *Evaluation/Earth Science.* Only Choice (2) provides a logical explanation that is consistent with the figure. The continents were once joined, and communities of organisms were spread across the entire supercontinent. Choices (1) and (5) are false; Choices (3) and (5) do not explain the situation that is described.
27. (2) *Application/Physics.* Heat is a form of energy, and the passage states that energy always flows from high-energy substances to low-energy substances. Cold water has less energy than a hot pan does, so heat will move from the pan to the water.
28. (2) *Evaluation/Biology.* Choices (3) and (5) are not likely to work in the long run if acid rain is not eliminated. Choices (1) and (4) are impractical. The only reasonable alternative is Choice (2).
29. (2) *Application/Physics.* The passage states that if one electrical device in a series circuit is disconnected, all the other devices will go out because the circuit has been broken. Choices (1) and (4) are false. Choices (3) and (5) are not relevant to the material. Choice (1) is the incorrect alternative to Choice (2).
30. (1) *Comprehension/Physics.* The passage clearly states that electrons flow from the negative pole to the positive pole.
31. (4) *Analysis/Biology.* The graphs show that the grasshopper population increases and decreases just after the plant population, its food supply. The last points on the plant graph indicate that the plant population is still declining. One would predict, then, that the grasshopper population also will continue to decline.
32. (5) *Evaluation/Biology.* The graph indicates a relationship between the amount of plant life and the number of grasshoppers in a given ecosystem. The most reasonable conclusion is that the two factors influence each other—probably that an abundance of plant life makes it possible for more adult grasshoppers to survive and reproduce, but an abundance of grasshoppers diminishes the amount of plant life, the food supply. Choices (2) and (3) introduce possibilities that are not covered in the graph. Choices (1) and (4) argue on the basis of somewhat illogical assumptions.
33. (2) *Application/Biology.* The graph indicates that over a long period of time, a decrease in food supply will be followed by a decrease in animal life. Choices (1), (3), (4), and (5) may help to decrease the mouse population temporarily, but as long as food is available, mice will survive to reproduce. Eliminating the availability of food is the best long-range solution.
34. (3) *Analysis/Chemistry.* If the reaction were reversible, simply adding water should result in the formation of the original substance, crystalline gypsum.
35. (2) *Comprehension/Biology.* The first vertebrates—animals with backbones—were fish. According to the passage, fish first appeared in the fossil record in the Paleozoic Era.
36. (1) *Analysis/Biology.* Forms of life during the Precambrian Era were simple and soft bodied. There would be no bones or shells to leave an imprint in what would become fossil rock. Fossils from this period of time, therefore, would be rare.
37. (2) *Comprehension/Biology.* According to this passage, much of North America was underwater during the Paleozoic Era. After that era, however, the oceans receded and more shoreline was exposed. Fossils of ocean life in the rocks must date from the Paleozoic Era, when that region was underwater.
38. (3) *Evaluation/Biology.* One trend that can be seen when reading the passage is that new groups of animals will appear on Earth eventually.
39. (1) *Analysis/Biology.* The highest oxygen-to-carbon-dioxide ratio occurs as the blood travels from the heart to the arteries after leaving the lungs. Choices (2), (3), (4), and (5) are false because after the blood enters the muscle cells, it exchanges oxygen for carbon dioxide.
40. (5) *Evaluation/Biology.* A species is better off if the genes of the individuals are mixed in the offspring: the offspring are healthier, and the chance of survival for the species is greater.
41. (3) *Comprehension/Biology.* The answer can be determined by reading the chart. Type B can receive blood from any type B or type O.
42. (4) *Application/Biology.* Type O blood can be donated to any other type; therefore, it is the most useful in a blood bank.
43. (2) *Analysis/Biology.* Because the father's genes must be OO, the child will inherit an O gene from him. The mother's genes may be AA or AO. If they are AA, the child will inherit an A from her. The child will be AO, or blood type A. If the mother is AO, she will pass either an A or an O gene to the child, whose blood type then will be either A (AO) or O (OO).
44. (1) *Evaluation/Biology.* Because type O is most common, it is seldom in short supply. None of the other choices explain the statement. Choice (4) is false, and Choices (2), (3), and (5) are not relevant to the question.
45. (2) *Evaluation/Biology.* One must assume that the part of the population that is being studied reflects the population as a whole

and is not unusual in any respect. Choice (3) would be impossible to meet. The other choices should not affect the results.

46. (3) *Application/Earth Science.* The passage explains that most of the moisture in the air is deposited just to the west of the mountains as the air rises and cools off.
47. (1) *Application/Earth Science.* Of the choices given, Choices (1) and (2) cover the longest distance and the same distance. Because the flight that is described in Choice (1) would require the plane to fly with a head wind, it would be a longer flight.
48. (4) *Comprehension/Physics.* The graph shows that after about 6000 years, half of the original 10 grams of carbon-14 is gone.
49. (3) *Comprehension/Biology.* The passage states that nerve cells coordinate the body's activities. These activities would include movements of muscles.
50. (4) *Analysis/Biology.* The passage notes that iron in red blood cells is involved in carrying oxygen to other cells of the body. An anemic person's red blood cells would not be able to deliver as much oxygen.
51. (1) *Evaluation/Biology.* All living things are made of cells. The description of a virus indicates that it is not made of cells and so would not be considered to be living.
52. (2) *Comprehension/Biology.* Choices (1), (3), (4), and (5) are false. Organs have specific functions in multicellular organisms; organelles serve specific functions in individual cells.
53. (5) *Evaluation/Biology.* The description notes nothing about the origins of bees or of plants. While it might be inferred, it is not explicit from the material. Therefore, Choice (5) is the correct answer.
54. (1) *Analysis/Earth Science.* Lines of longitude cross at the poles of Earth. When this crossing is disrupted by the flattening of the globe, the lines of longitude become parallel. Land masses near the poles are distorted in the process.
55. (3) *Application/Earth Science.* The lines of longitude and latitude divide the surface of Earth into an imaginary grid. This grid gives sailors an absolute reference for the description of locations on the sea. All other choices are not relevant.
56. (5) *Comprehension/Chemistry.* Because acids have a pH that is lower than 7 and bases have a pH that is higher than 7, it is clear that adding a base to an acid will increase the pH of the acid and will bring it closer to pH 7 (neutral).
57. (2) *Application/Chemistry.* The first paragraph explains how the pH scale measures acidity. Acid rain is 100 times more acidic than normal rain (pH 6); therefore, it must have a pH of 4.
58. (3) *Analysis/Chemistry.* The base should just neutralize the acid and result in a saltwater solution that has a pH of 7.
59. (3) *Analysis/Chemistry.* The passage notes that blueberries grow well in acidic soil and that evergreen needles increase soil acidity.
60. (4) *Analysis/Biology.* The proteins that cause clotting are found on the blood cells. If the patient is given only plasma, his or her blood should not clot.
61. (3) *Comprehension/Physics.* Reading across to the Celsius scale from 0 degrees on the Fahrenheit scale gives a value of about $^{-}20$ degrees Celsius.
62. (3) *Analysis/Physics.* The two scales use the same interval of measurement; the difference between them is found in their designation of a point of 0 degrees. The Kelvin scale uses absolute zero; the Celsius scale uses a point that is 273° above absolute zero.
63. (4) *Analysis/Physics.* The figure states that °K = °C + 273. Because water boils at 100°C, the boiling point of water on the Kelvin scale is 100 + 273, or 373°.
64. (5) *Analysis/Physics.* Loudness is due to amplitude, not frequency (thus eliminating Choices [1] and [2]). Nothing in the passage discusses the role of air molecules (eliminating Choice [3]). A decreasing amplitude would result in decreased volume (eliminating Choice [4]). The amplitude will decrease over distance, but extra energy at the source will sustain the amplitude for a fair distance.
65. (4) *Evaluation/Physics.* The amount of energy in the source affects the amplitude (loudness), not the frequency (pitch). No matter how hard the C tuning fork is struck, its pitch will not change. Choices (1), (3), and (5) are false. Choice (2) is true, but it is not illustrated in the experiment.
66. (3) *Analysis/Earth Science.* As the North Pole tilts toward the sun, the South Pole will tilt away from the sun. Therefore, as the days grow longer in the Northern Hemisphere, they will grow shorter in the Southern Hemisphere.

SIMULATION
Simulated Test

1. (4) *Comprehension/Physics.* Because the mass of the object is slightly less after the object gives up energy, it is reasonable to conclude that a small amount of mass is lost in the energy-releasing process shown in the diagram.
2. (5) *Comprehension/Earth Science.* Because salts are denser than water, seawater also must be denser than distilled water. As the proportion of salt increases, so will the density (which eliminates Choices [2] and [3]). The information in Choices (1) and (4) is not discussed in the passage.
3. (3) *Application/Biology.* The specialized cells of higher plants make *parenchyma* a kind of plant tissue.
4. (4) *Application/Biology.* Because the skin is made up of several tissues (groups of specialized cells that serve a common function), skin is classified as an organ.
5. (5) *Application/Biology.* All of the structures named are organs of the nervous system, which is an organ system.
6. (1) *Application/Biology.* A cell is defined as the simplest unit that has all the characteristics of a living organism. A virus has only one of the characteristics of life—the ability to reproduce, so it is not a cell, eliminating Choice [2]. (Choices (3), (4), and (5) are various combinations of cells.)
7. (2) *Application/Biology.* The lack of specialized cells indicates that no tissues exist in the brown algae; thus, no organs (Choice [4]) or organ systems (Choice [5]) can exist. The highest form of organization is that of the cell.
8. (5) *Analysis/Chemistry.* Warm temperatures are required for fermentation; therefore, refrigerating the cider would slow the process.
9. (2) *Comprehension/Biology.* The great length of the intestine, the folding of the brain and the intestine, and the large number of very tiny balloonlike sacs of the lung each provides a large surface area for more efficient exchange of materials. The adaptations described increase the size and the weight of the organ (which eliminates Choices [1] and [4]). The passage does not refer to speed (Choice [3]) or mobility (Choice [5]) as adaptations.
10. (4) *Comprehension/Earth Science.* A front is formed at a point where two air masses meet. The figure, which depicts the formation of a warm front, shows a warm-air mass that is moving into contact with a stationary cold-air mass that is not moving.
11. (1) *Application/Earth Science.* As warm air rises over cold air, its temperature drops. As it cools, its ability to hold moisture decreases. The result will be rain.
12. (3) *Evaluation/Earth Science.* Just as rain will fall when warm air rises above cold air in the atmosphere, so water vapor will condense when warm air rises above colder air in a room. Choices (1), (2), and (4) could be true, but they are not based on information in the map. Choice (5) is false.
13. (4) *Comprehension/Biology.* The passage does not address specifically at which point the digestive system can first be seen; however, it notes that all systems are visible by eight weeks.
14. (3) *Application/Biology.* Both extensors and flexors are needed to lift food and then to return to the food source, and to chew food (which eliminates Choices [1] and [2], for either choice is incomplete by itself). The cardiac muscles (Choice [4]) and involuntary muscles (Choice [5]) are not mentioned in the passage.
15. (1) *Comprehension/Physics.* The passage states that any two objects that are thrown from the same height will strike the earth at the same time—an indication that gravity pulls on each one. The acceleration that is due to gravity is a constant.
16. (3) *Comprehension/Physics.* Friction normally slows a thrown object as it travels through the air. A constant horizontal speed assumes the absence of friction. Vertical speed, which is affected by gravity, is not involved (which eliminates Choices [1] and [2]). It is not necessary to assume anything about the weight of the object (Choice [4]) or about the absence of gravity (Choice [5]).
17. (4) *Evaluation/Physics.* Because gravity begins to pull an object downward as soon as it is thrown, the ball that was thrown in an arc would remain in the air slightly longer than the ball that was thrown horizontally. Thus, there would be enough time for the ball to travel to second base.
18. (4) *Evaluation/Biology.* According to the passage, there are more plants than herbivores and more herbivores than carnivores. If the food supply is limited, it would be logical to eat fewer of the less-abundant organisms and to eat more of the most readily available food sources. A plant-based diet, therefore, would make sense. Choices (1) and (3) would deplete the meat supply. Choice (2) is not relevant to the passage, and Choice (5), while possible in theory, is not practical.
19. (4) *Analysis/Chemistry.* As long as water remains unpolluted, the effect of minerals on its taste is a matter of opinion. Choices (1), (2), (3), and (5) are based on the chemical properties of hard water and soft water.
20. (2) *Analysis/Biology.* Lactase is the enzyme that is responsible for digesting lactose. People who are "lactose intolerant" cannot produce enough lactase to digest the lactose in the milk products they consume.
21. (4) *Application/Biology.* Because vitamin B_1 contributes to the functioning of nerves and muscles, and because vitamin D contributes to the development of bones, these vitamins would be especially important to athletes.
22. (3) *Application/Biology.* Surgery would require the healing of an

incision and would involve some loss of blood. Therefore, vitamins that aid in healing (C), in the forming of blood (E), and in the clotting of blood (K) could prove to be helpful. Taking massive doses of vitamin K (Choice [1]) is unwise, for high levels of fat-soluble vitamins can be toxic. Vitamin B_6, which strengthens the immune system (Choice [4]), could be helpful, but massive doses are worthless because the excess will be excreted. Choice (2) could be harmful if the patient tends to be deficient in vitamin C. Choice (5) includes vitamins A and D, whose applications probably would not extend to surgery.

23. (2) *Analysis/Biology.* Cod liver oil, itself a fat, can facilitate the absorption of fat-soluble vitamins.
24. (2) *Analysis/Biology.* Because vitamins are directly related to the formation of enzymes, a lack of vitamins could result in the slowing down of chemical reactions in the body. The body cannot produce vitamins itself (Choice [4]). There is no evidence in the passage to suggest Choices (1), (3), or (5).
25. (3) *Evaluation/Biology.* The passage does not discuss which vitamins are more likely to be deficient. Choice (1) is supported by the passage, which states that excessive amounts of fat-soluble vitamins can be toxic. Choice (2) is supported, for the passage states that excess amounts of water-soluble vitamins (such as vitamin C) will be excreted. Choice (4) is supported by the mention of recommended daily dosages. Choice (5) is supported by the mention that enzyme formation is directly linked to the presence of vitamins.
26. (5) *Evaluation/Earth Science.* Because a large body of water can absorb or release large amounts of heat without its own temperature changing, the temperature of the surrounding area is moderated. Thus, location E would be cool in the summer and warm in the winter compared to inland areas A and B.
27. (3) *Analysis/Physics.* Inertia is the force that resists change in motion. On a forward-moving bus, you, too, are moving forward. If the bus stops short, you will continue to move forward.
28. (2) *Analysis/Chemistry.* Because magnesium is ranked above zinc in the activity series, it will replace the zinc in the lining of the bucket.
29. (4) *Analysis/Chemistry.* Platinum ranks very low in the activity series. It is unreactive and thus is replaced in a compound easily. Therefore, platinum is the metal that is most likely to be found free and uncombined in nature.
30. (3) *Analysis/Biology.* The nerve pathways through the body include messages that are being relayed from the brain to the muscles. Choices (1) and (4) are improbable. If there is enough heat to burn the skin, there is enough heat to stimulate the nerves (Choice [2]). Choice (5) is incorrect because it is the sensory nerves, not the motor nerves, that send messages to the brain.
31. (4) *Analysis/Chemistry.* Because liquid is the state of matter that exists between solid and gas, it can be assumed that the molecules of a liquid would have more freedom of movement than would the molecules of a solid but less freedom of movement than would the molecules of a gas.
32. (2) *Comprehension/Earth Science.* According to the passage, transported soils are formed by particles that are carried along and then deposited by running water.
33. (4) *Application/Earth Science.* Coarse grains of sand do not pack closely; thus, they have open spaces between them. As a result, water runs off easily, and air can reach the roots.
34. (1) *Application/Earth Science.* Because fine clay packs densely, water does not drain quickly. A clay-based soil, therefore, would be the most water-retentive soil.
35. (5) *Analysis/Earth Science.* Soil that has been covered by a building probably has had few or no organisms dying and decaying in it. It will be lacking in the organic nutrients that come from this source.
36. (1) *Evaluation/Earth Science.* The microorganisms that are responsible for the decay that results in organic nutrients flourish in warm, moist climates. Of the choices given, the climate of the Gulf Coast is the warmest and the most moist.
37. (3) *Application/Chemistry.* The liquid water has become a gas, slowly, without boiling.
38 (2) *Application/Chemistry.* The rapid change of a liquid to a gas is a definition of *boiling*.
39. (4) *Application/Chemistry.* The cooling of a gas to its liquid state is a definition of *condensation*.
40. (1) *Application/Chemistry.* Copper that can be poured is in the liquid state—a result of melting.
41. (5) *Application/Chemistry.* Salt crystals are solid. Cooling a liquid to form a solid is a definition of freezing.
42. (4) *Analysis/Biology.* According to the figure, the vascular tissue is responsible for the transport of nutrients and water. In animals, this is one of the primary functions of the circulatory system. None of the other plant tissues has a function that is similar to that of the circulatory system.
43. (4) *Analysis/Biology.* Sun and warmth increase the loss of water through the stomata. Opening the pores only at night would decrease the evaporation loss, for the sun would have set and the temperature would have dropped. Choices (1), (2), and (3) would make it impossible for the plant to survive; Choice (5) would result in more, not less, loss of water.
44 (1) *Comprehension/Biology.* According to the chart, only one male child of Victoria and Albert—Leopold—had hemophilia. Choices (2), (3) and (4) are contradicted by the chart. The chart provides no evidence for Choice (5).
45. (5) *Application/Biology.* A mother who has hemophilia must have two genes for the condition, or she would not manifest the disease. Certainly she will pass the gene on to her son. Because a man has only one X chromosome, which is inherited from his mother, the son will have a 100% chance of having hemophilia.
46. (2) *Analysis/Biology.* Because men interit sex-linked conditions from their mothers, the fact that Victoria's son Leopold had hemophilia means that she must have carried the gene.
47. (1) *Evaluation/Biology.* The chart shows that with each succeeding generation, the incidence of hemophilia increased. This trend is typical of the results of inbreeding

on recessive traits (and eliminates Choices [4] and [5], for hemophilia is transmitted only through inheritance). Neither having more children (Choice [2]) nor having more male children (Choice [3]) is a result of inbreeding.

48. (2) *Evaluation/Biology.* Because hemophilia runs in the woman's family, the woman may be a carrier even though she does not manifest the disease. She risks passing the disease on to her children, particularly if she marries a hemophiliac.

49. (1) *Application/Earth Science.* Dew forms most readily when there is much moisture in the air. Frost is dew that has frozen. Although frost could form when the air is cool and dry (Choice [3]), it is most likely that frost would form when the air is cool and damp. Warm air (Choice [2]) would not allow frost to form at all.

50. (3) *Comprehension/Physics.* The product of the force and the length of the lever arm on the left side of the fulcrum is $15 \times 4 = 60$. Because the product on the right also must equal 60, and because the lever arm is 6 feet long, the force must be 10 pounds: $15 \times 4 = 10 \times 6 = 60$.

51. (4) *Application/Biology.* Air that enters the respiratory tract at the trachea will bypass the pharynx and the larynx. A tracheotomy, therefore, can be used to provide for inhalation in the case of a blocked larynx.

52. (1) *Analysis/Biology.* Sound, such as talking, is produced as air passes over the larynx during exhalation. For this process to occur, the connection from the larynx to the pharynx must be open. This passage normally closes upon swallowing to prevent food from entering the larynx. If one talks and eats at the same time, this passage may remain open, and food may enter the respiratory tract, causing choking. None of the other choices involves the relationship between the digestive and the respiratory tracts.

53. (1) *Analysis/Biology.* For oxygen to enter the blood, the walls of the alveoli must remain moist and intact. Choices (2), (3), and (4) would not affect the amount of oxygen; Choice (5) would result in an even greater supply of oxygen.

54. (4) *Analysis/Biology.* The medication will be swallowed and will enter the digestive system from the throat. The respiratory tract, including the larynx, will not be affected.

55. (4) *Evaluation/Biology.* In a situation in which oxygen may be in short supply, people at risk should avoid activities that would increase their consumption of oxygen. They also should not spend excessive time outdoors, where air pollution is likely to be heavier. Temperature (Choice [2]) and humidity (Choice [1]) are not involved, exercise (Choice [3]) would only aggravate the problem, and muscle relaxants (Choice [5]) would have no effect.

56. (2) *Analysis/Earth Science.* Because warm water can hold less oxygen than cold water, the oxygen level in the river will go down each time the river is polluted by hot water. Because the river moves rapidly, however, the hot water will move on, and the decrease in oxygen will not be permanent.

57. (2) *Comprehension/Chemistry.* An atom of a radioactive element will continue to decay until an atom of a nonradioactive element is attained.

58. (1) *Application/Chemistry.* Because alpha particles have a positive charge, beta particles have a negative charge, and gamma rays have no charge, these three types of radiation can be separated by an electric field. Blocks of lead and concrete (Choice [3]) would stop all three forms of radiation and thus would not separate any of them. Sheets of paper (Choice [4]) would stop alpha particles but would not separate beta particles from gamma rays.

59. (4) *Analysis/Chemistry.* Control rods are needed to regulate the speed of a reaction. Their malfunction would encourage the reaction to run away with itself and produce large amounts of heat very quickly. The result easily could be an explosion. Choices (1), (2), and (5) are in direct opposition to the situation described, and there is no evidence to support Choice (3).

60. (2) *Evaluation/Chemistry.* Because the emission of radiation nearly always includes gamma rays, the radiation from nuclear wastes could penetrate any material that is less dense than lead or concrete. Choices (1) and (5) could be true, but they are not based on information in the passage. There is no evidence to support Choice (3). Choice (4) is not a realistic explanation because most radioactive wastes remain radioactive for millions of years.

61. (2) *Comprehension/Biology.* The table lists tetracycline, an antibiotic, as a treatment for acne.

62 (4) *Comprehension/Biology.* Of the two antibiotics listed, penicillin is the one that is used to treat pneumonia.

63. (5) *Analysis/Biology.* The description of alcohol indicates that it is a depressant. Sleeping pills are depressants, as well. The combination of two substances that slow vital functions can be fatal.

64. (5) *Evaluation/Biology.* As a patient recovers from a viral infection, such as influenza, the immune system produces antibodies which give immunity to the virus that caused the infection. Choices (1) and (4) are true of viruses, not all communicable diseases. Choices (2) and (3) are false.

65. (4) *Analysis/Physics.* Air pressure is measured by the height of mercury in the tube of the barometer. Air presses down on the reservoir of mercury, causing the mercury to be forced up the tube. (The greater the air pressure is, the greater the force will be on the mercury in the tube, and the higher the mercury in the tube will rise.)

66. (3) *Comprehension/Physics.* According to the figure, a second-class lever may require lifting, a downward motion, or a pushing-together motion.